Behavioral and Metabolic Aspects of Breastfeeding

International Trends

Volume Editors

A.P. Simopoulos
The Center for Genetics, Nutrition and Health,
Washington, D.C.

J.E. Dutra de Oliveira
University of Ribeirao Preto,
São Paulo

I.D. Desai
University of British Columbia,
Vancouver, B.C.

22 figures and 52 tables, 1995

KARGER

Basel · Freiburg · Paris · London · New York ·
New Delhi · Bangkok · Singapore · Tokyo · Sydney

World Review of Nutrition and Dietetics

Library of Congress Cataloging-in-Publication Data
Behavioral and metabolic aspects of breastfeeding: international trends / volume
editors, A.P. Simopoulos, J.E. Dutra de Oliveira, I. Desai.
(World review of nutrition and dietetics; vol. 78)
Includes bibliographical references and index.
1. Breast feeding.   2. Lactation – Nutritional aspects.   3. Infants – Nutrition.   4. Breast feeding –
Social aspects.   I. Simopoulos, Artemis P., 1933–   .   II. Oliveira, José E. Dutra de.   III. Desai, Indrajit
Dayalji.   IV. Series.
[DNLM: 1. Breast Feeding. 2. Lactation – physiology. 3. Weaning. 4. Milk, Human. 5. Infant
Nutrition.   W1 W0898 v.78 1995 / WS 125 B419 1995] QP141.A1W59 vol. 78 [RJ216]
612.3 s – dc20 [613.2'6]
ISBN 3–8055–6102–4 (alk. paper)

Bibliographic Indices. This publication is listed in bibliographic services, including Current Contents® and Index Medicus.

© Copyright 1995 by S. Karger AG, P.O. Box, CH–4009 Basel (Switzerland)
Printed in Switzerland on acid-free paper by Thür AG Offsetdruck, Pratteln
ISBN 3–8055–6102–4

....
Behavioral and Metabolic Aspects of Breastfeeding

World Review of Nutrition and Dietetics

Vol. 78

Basel · Freiburg · Paris · London · New York ·
New Delhi · Bangkok · Singapore · Tokyo · Sydney

....

Contents

**Breastfeeding for Optimal Development. The Alpha and the
Omega in Human Milk**

....

Preface

That breastfeeding is the preferred form of infant feeding has been stated both by professional societies and international organizations. Since the Second World War, research on the composition of human milk has generated data defining its protein and amino acid, fat, carbohydrate, vitamin and mineral content. Infant formula was developed by manufacturers to resemble the known composition of human milk. A number of studies have been carried out on maternal-infant interaction and bonding comparing breastfed versus bottle-fed infants, taking into account maternal education and socioeconomic status. Over the past 10 years, a revolution has occurred in studies on the fat composition of human milk. Fat in human milk not only provides a source of calories as originally thought, but its composition of fatty acids differs significantly from that of the milk of cows and other sources used in infant formula. Human milk is rich in essential fatty acids (EFAs), both ω6 and ω3 polyunsaturated fatty acids (PUFAs). EFAs, particularly docosahexaenoic acid (DHA), play important roles in the normal function and development of the retina and brain. Furthermore, the premature infant not only has relatively smaller amounts of DHA, but its overall nutritional needs in terms of other nutrients and energy are also more than those of full-term infants. Therefore, human milk from mothers of full-term infants needs to be appropriately modified to suit the needs of the premature infant.

There is a resurgence and expansion of research on infant feeding, from behavioral (mental development) as well as biochemical perspectives. Emphasis is given to the role of EFAs in learning and behavior, since comparisons between infants fed formulas containing DHA versus those that did not suggest that the latter have delay in visual function and significant differences in intelligence quotient at 8 years of age.

It is now clear that human milk is uniquely adapted to meet the biochemical and functional needs of infants, yet the rate of breastfeeding is decreasing throughout the world. Only the actual percentages differ from country to country. This volume therefore includes descriptions of programs to improve breastfeeding rates and international activities to increase breastfeeding. 1994 will be remembered as the year when the United States voted 'yes' to support the International Code for Marketing of Breast-Milk Substitutes. This volume provides the scientific evidence to justify breastfeeding, an overview of its status in selected countries around the world, and the programs that aim to increase the rate of breastfeeding. Overall, this volume of *World Review of Nutrition and Dietetics* provides the reasons and the techniques for the promotion of breastfeeding.

The volume begins with the statement 'Recommendations for the Essential Fatty Acid Requirement for Infant Formulas' developed by the Board of Directors of the International Society for the Study of Fatty Acids and Lipids, which clearly states the need to supplement infant formula with appropriate amounts of EFAs. Three papers on the behavioral and metabolic aspects of breastfeeding follow. The first paper in this section, 'Breastfeeding for Optimal Mental Development: The Alpha and the Omega in Human Milk' is by Isidora de Andraca and Ricardo Uauy. Dr. Uauy and his group have contributed enormously to our knowledge on the role of EFAs and particularly DHA in the function of the retina and cerebral cortex of the premature and full-term infant. Their paper represents a critical review of studies that compare developmental outcomes of breastfed and formula-fed infants, mother-infant interactions, and the evidence as to how breastfeeding may affect them. They evaluate the methods used in determining early interactions and psychoaffective bonding, the breastfeeding act as a unique scenario for bonding and attachment, and the effect of feeding type on developmental outcomes. The paper concludes with a precise review and critique of EFAs as determinants of retinal and brain function. The authors state that 'The challenge faced by future studies will be to establish a conceptual framework where key mechanisms and responses may be unraveled. An integrated approach is needed to assess the nonnutritional consequences of breastfeeding. The design of such studies should consider assessing not only the effect of psychosocial factors affecting development but also the subtle interactions between mother, infant and the environment. Maternal psychological makeup, family interactions, and the environmental stimulation the child receives are key factors that should not be neglected. Case-control studies represent the best next step in this fascinating field of biological research'.

The second paper on 'Breastfeeding Kinetics' by Verity Livingstone presents a problem-solving approach to breastfeeding difficulties. This paper covers subjects such as mammogenesis, lactogenesis and galactopoiesis, and considers factors that help or interfere with these processes. The section on milk transfer discusses the effects of positioning, attachment suckling, and the factors which impede milk transfer. The section on milk intake considers the frequency, duration, and pattern of breast use and the factors that help or impair milk intake. The last three sections include descriptions of the insufficient-milk syndrome, and infant and maternal hyperlactation syndromes.

'Human Milk and Premature Infants' by F.E. Martinez and I.D. Desai points out that the nutritional requirements of premature infants are greater than those of term infants. In fact, the nutritional requirements of the premature are greater than at any time in human life, since at no other stage of extrauterine life does body weight triple within 3 months. The authors discuss the importance of human milk for premature infants, the merits of human milk versus special formulas, homogenization of human milk, clinical trials with ultrasonically homogenized human milk, human milk fortifiers, and recommend a formula based on human milk that is more appropriate for premature infants.

The final part of the book looks at international trends and actions in selected countries. Because the rate of breastfeeding is decreasing, governments are developing programs to reverse this trend. Breastfeeding is essential for the normal growth and development of all children in the world, and these last chapters describe breastfeeding patterns in selected countries around the world.

The paper 'Breastfeeding in Australia' by Margaret Lund-Adams and Peter Heywood, is an extensive presentation of the breastfeeding trends among Aborigines and Europeans, the role of ethnicity, socioeconomic status, and reasons given for stopping breastfeeding, the support for breastfeeding provided by international initiatives and the government, national goals and targets, and their monitoring. This paper discusses the International Code of Marketing of Breast-Milk Substitutes, monitoring of the World Health Organization (WHO) Code, the World Summit, the Baby Friendly Hospital Initiative, and support provided by nongovernment organizations.

'Breastfeeding in Korea' by Sook He Kim, Kyoung Kim Woo, Ae Lee Kyoung and Young Oh Se, provides information on infant feeding practices in Korea and the factors affecting these practices. The authors conclude that the changes in infant feeding practices are due to Westernization, and improvement in women's education with the result that fewer women breastfeed.

In 'Breastfeeding in China', Dong-Sheng Liu and Xibin Wang describe trends in breastfeeding and provide data on the composition of human milk in China. As in other countries, breastfeeding rates are decreasing and the government has developed programs to increase it.

'Breastfeeding Trends in Cuba' by Manuel Amador, Luis C. Silva, Graciela Uriburu, Marta Otaduy, and Francisco Valdes, provides an extensive review of the National Survey and the differences in the prevalence of breastfeeding in the various provinces, and between urban and rural areas. The authors provide extensive descriptions of the government's plan, including strategies and actions to increase breastfeeding.

The final paper 'Breastfeeding Patterns in the Arabian Gulf Countries' by Abdulrahman O. Musaiger, describes breastfeeding and weaning practices in the Gulf countries (Bahrain, Kuwait, Oman, Qatar, Saudi Arabia, and the United Arab Emirates). Activities to encourage and support breastfeeding in the Gulf consist of actions taken to improve maternity protection, educational activities, marketing activities, the support of appropriate weaning practices, workshop and seminar activities, and research.

The publication of this volume is a joint effort of the three editors: Artemis P. Simopoulos, J. E. Dutra de Oliveira, Past President of the International Union of Nutritional Sciences, and his coworker, Dr. I.D. Desai, and the contributors are scientists from Australia, Brazil, Canada, Chile, China, Cuba, Korea, and the United Arab Emirates.

The volume should be of interest to researchers, biochemists, nutritionists, midwives, dietitians, behavioral scientists, physicians, industry scientists, administrators, policy makers, and specialists in international health and development.

Artemis P. Simopoulos, MD

....

ISSFAL Board Statement

**Recommendations for the Essential Fatty Acid Requirement for
Infant Formulas (June 1994)**

Background

Human milk is the best and only time-proven source of fat and essential fatty
acids (EFAs) in the infant diet. It has long been understood that ω6 fatty acids are
essential for proper growth and development of mammals and must be included in
infant formula. In addition, there is now strong experimental and clinical evidence
indicating that dietary ω3 fatty acids are essential for normal nervous system
development in the human. In preterm infants, consumption of formulas based
on, for example, corn oil not only induces biochemical signs of ω3 deficiency but
also affects retinal responses to light, cortical visual evoked potentials and
behavioral measures of visual development.

Preterm infants also require docosahexaenoic acid (DHA) because they are
unable to form a sufficient quantity from precursors provided by vegetable-oil-
based formula products. Preterm infants fed soy-oil-supplemented formulas have
altered electroretinograms and delayed visual maturation relative to those fed
DHA-supplemented formula or human milk. These changes are correlated with
biochemical indices of low DHA status. In addition, preliminary evidence suggests
that cognitive scores may also be affected by low dietary DHA supply.

Proper formulations (see below) containing marine oils have been safely used
in formula, i.e., without adverse effects on growth, mental development, bleeding
time and lipid peroxidation. Sources of long-chain ω3 and ω6 polyunsaturated
fatty acids that are currently being tested include lipids derived from marine
sources, egg or tissue phospholipids and those from algae and fungi; these latter
organisms have been selected to produce oils enriched in DHA or arachidonic acid

(AA). The safety of each source of long-chain polyunsaturates must be fully evaluated before they are added to infant formula.

Specific Recommendations for Formula-Fed Infants

There are three possible approaches in defining recommendations for formula-fed infants. The first is to mimic the provision of fatty acids provided by human milk from well-nourished women. The second is to replicate what has been tested and found efficacious and safe in clinical trials of long-chain polyunsaturated fatty acid supplementation. The third is to provide sufficient EFAs to meet the accretion needs of late fetal and early postnatal life; the data for this latter approach are somewhat limited.

The results obtained with the first two approaches are in general agreement with each other. Since there is variability in the composition of human milk and in what has been clinically tested, we propose acceptable ranges for EFAs rather than fixed target figures.

Since we recognize not only $\omega 6$ but also $\omega 3$ fatty acids as essential nutrients for premature infants, the following provisional recommendations are suggested for the formula-fed preterm infant for the first 6 months of life.

(1) The total EFA requirement ($\omega 6 + \omega 3$) for premature infants should be set at 5–6% of total energy up to a maximum of 13%. This represents approximately 0.6–0.8 g/kg/day with an upper limit of 1.5 g/kg. The total linoleic acid may adversely affect the formation of long-chain polyunsaturates.

(2) The recommended range for the parent $\omega 6$ EFA (linoleic acid) supplied should be 0.5–0.7 g/kg daily. Since fatty acid biosynthetic enzyme activities may be limited in the premature infant, formulas for these infants should provide 60–100 mg/kg/day as preformed AA and its associated long-chain $\omega 6$ forms ($22 : 4\omega 6$ and $22 : 5\omega 6$).

(3) The total $\omega 3$ fatty acid supply should be 70–150 mg/kg daily. Since desaturase and elongase enzyme activity may be limited in premature infants, formulas for their use should provide 35–75 mg (i.e., 50%) of the $\omega 3$ fatty acid per kilogram body weight per day as DHA.

(4) The ratio of total $\omega 6$ to $\omega 3$ fatty acid present in the early diet should be maintained within a range of 5 : 1 to a maximum of 10 : 1. The ratio of DHA to AA should be in the range of 1 : 1 to 1 : 3. The ratio of DHA to eicosapentaenoic acid should be about 5 : 1 or higher.

Despite the body of evidence that has recently accumulated for human infants, the American Academy of Pediatrics USA has still not acknowledged the need for and the essentiality of $\omega 3$ fatty acids; even today, low linolenic acid formulas are still in use in some parts of the world. The European Society for

---

Pediatric Gastroenterology and Nutrition in 1991, the British Nutrition Foundation in 1992 and the WHO/FAO Expert Committee on Fats and Oils in Human Nutrition which met in 1993 recommended not only that linolenic acid be present but that DHA and AA should also be added to formulas destined for preterm infants. In general, the specific recommendations given above are applicable to term infants; however, the requirements for long-chain polyunsaturates (AA and DHA) for term infants await the results of clinical trials that are now in progress.

Simopoulos AP, Dutra de Oliveira JE, Desai ID (eds): Behavioral and Metabolic
Aspects of Breastfeeding. World Rev Nutr Diet. Basel, Karger, 1995, vol 78, pp 1–27

....

Breastfeeding for Optimal Mental Development

The Alpha and the Omega in Human Milk[1]

Isidora de Andraca, Ricardo Uauy

Instituto de Nutrición y Tecnología de los Alimentos (INTA), Universidad de
Chile, Santiago, Chile

Contents

Introduction

A critical review of studies comparing the developmental outcomes of
breastfed and formula-fed infants is complex because there are multiple differences between groups of infants that have been fed by breast or bottle that, for
obvious reasons, cannot be controlled. The ideal way to address this question

[1] Supported in part by Fondecyt grants no. 1930820 and no. 1950241.

would be a prospective randomized study controlling not only for what is fed, but also for who feeds it, and for how it is fed. Since, for practical and ethical reasons, this cannot be done, we are left with results from mostly retrospective, and a few prospective studies, where some, but not all, intervening variables have been considered in the evaluation.

Formula differs from human milk in several nutritional components which may affect the biological basis of mental function. There are also specific characteristics associated with parents who choose breast milk over formula feeding of their infants. The decision whether or not to breastfeed is associated with parental and infant factors which will confound comparisons of breast- and bottle-fed infant groups. There is also a potential effect of taking the milk from the breast. The act of breastfeeding itself has profound effects on the behavior and physiology of mother and infant. These factors may all play a role in developmental outcomes affecting the relationship between type of feeding and mental function.

This paper will review mother-infant interaction and the evidence as to how breastfeeding may affect it, the methods to evaluate the effect of the type of feeding on mental development, and the results of the available studies that have addressed this issue. Finally, we will review the exciting recent findings of investigations which have examined the presence of ω3 fatty acids in human milk or in the early diet as determinants of retinal and brain development.

Breastfeeding as a Determinant of Mother-Infant Interaction

Breastfeeding has been the traditional time-honored method to feed neonates and infants. Artificial feeding first became progressively popular in urban developed societies, then gained widespread use in urbanized developing societies in the last few decades. The advantages of breastfeeding were questioned in the past, but recently mother's milk has regained its reputation for being best. The renewed interest in breastfeeding as the best feeding method for babies has been conditioned in part by new scientific knowledge on the biological basis for the perceived benefits. The potential advantages of breastfeeding on mental function and social and emotional development have not been explicitly recognized in breastfeeding promotion efforts. This section will review the evidence for a breastfeeding effect on the social and emotional development of infants and its relationship with mother-infant bonding.

Early Interactions and Psychoaffective Bonding

It is generally accepted that all human beings need nutritional and psychosocial support for growth and development. These basic human needs must be fulfilled if normal psychological and somatic development are to occur. The early

---

interactions between infants and parents, especially mothers, are crucial for defining the type of relationship the infant will establish with his/her environment (mainly with other people) as an infant and later as an adult. Studies of infants who have been separated from their biological mothers early in life document that the quality of maternal-infant bonding is determined by these early contacts and that this phenomenon has implications for emotional development and adult psychoaffective behavior [1].

Klaus and Kennel [2] have clearly demonstrated that bonding between parents and children is initiated early – from the time the decision to have a child is made. After conception, bonding is strengthened as the fetus develops; maternal perception of fetal heart sounds, movements and real sensory input at the time of the prenatal health visit, all contribute to the establishment of bonding before birth has occurred. The mother may also fantasize about how the infant is developing or what his or her future will be like. Listening to a fetal heart beating, and watching echocardiographic images of fetal movements are small examples of sensory input which support the bonding process.

The moment of delivery and the first postpartum hours provide a special opportunity to support strong bonding by parents which will result in good infant attachment. During the initial day, the infant will characteristically be in a hyperalert state, unless the mother has been excessively sedated. This will make the child more responsive to sensory stimuli, especially those from the mother. Infants are able to discern the high pitch of the maternal voice and will selectively respond to that stimulus. They will recognize parents' voices because in utero exposure has occurred and will orient their attention and move rhythmically in response to them [2, 3].

Visual contact is also evident during this period and helps to establish attachment. Infant behavior including crying, rooting, sucking, startling and other reflex responses all contribute to the infant's adaptation to his/her new surroundings. These interactive responses facilitate bonding and as a consequence generate better infant care by mothers. Such behavioral phenomena provide an important selective advantage given the high vulnerability of the human neonate [3].

The mother also experiences important behavioral changes conditioned in part by hormone changes which occur after the placenta has been delivered. Her sensitivity to infant cues becomes enhanced and she adapts her responses to the infants' sensory capacities. The mother will modify her tone of voice making it more audible, she aligns her gaze with that of the infant and modifies her rhythm to make it interdependent with the infant's. This period of enhanced maternal sensitivity lasts approximately 3 days during which she is especially responsive to the infant's needs. During this period, the reciprocal attraction determines complementary behavioral responses which have been characterized as 'falling in love'. A stimulus by one elicits a behavioral response by the other, which reinforces and satisfies the originator. Thus, through progressive positive reinforcement, bonding is established: neither child nor mother will ever be the same again. For example, if the infant cries, the mother will cuddle him/her; the proximity of the mother induces tranquility in the infant. This signals to the mother her importance to the infant, reinforcing a caring behavior. The responses are not a chain of events but rather a sequence in which a specific behavior triggers a set of responses by the other member of the dyad. The set of interrelated behavioral responses by the mother and infant dyad is what constitutes the basis for bonding [3, 4].

Early, close and prolonged skin-to-skin contact between mother and infant constitute another influence which helps to establish and strengthen the bonding process. A study by Kennel et al. [5] indicated that when mothers were given 15 additional minutes of close contact with their infants in the first 30 minutes of postnatal life, they exhibited an enhanced affect towards their infant. Those that had the extra early close contact spent more time looking at and kissing the infant; control mothers spent a similar amount of time with the child, but occupied in general care and in hygienic practices. Infants from the early contact group were less irritable, cried less and smiled more. Other studies have demonstrated that early postpartum skin-to-skin contact is associated with a higher prevalence and greater duration of breastfeeding.

Recent studies indicate that physical contact has positive effects not only during the immediate neonatal period but later in infancy as well. The early contact with the mother's chest provided by soft baby carriers influences the quality of the infant attachment process and the bonding the mother establishes with that infant. This promotes maternal-infant interactions which favor psychosocial and emotional development. Thus the specific situation of early contact between mother's breast and infant's mouth and face cannot be reproduced in the bottle-fed infant [6].

Maternal-infant interactions are set in motion very early and are subject to multiple factors. Klaus and Kennel [2, 3] have differentiated between stable factors which cannot be modified and those that are modifiable by cultural, social and health practices. The mother's genetic background, early childhood antecedents, marital experiences and pregnancy conditions are examples of the former. Factors that can be modified and have a profound impact on maternal-infant interactions relate to hospital practices, the message provided by health professionals, and early contact or separation provided during the first hours and days after delivery. It is in this area where specific intervention initiatives can be undertaken to promote bonding and attachment. Many actions are being promoted today in this direction: the UNICEF friendly hospital initiative is a prime example. Avoidance of early separation of mother and infant and the facilitation of early contact right after birth are examples of what can be done. Parents will respond favorably to a demonstration of the infant's cognitive capacities and the role of early reflexes in adaptation to the new environment. Parental recognition of the infant's sensibility and of the potential consequences of their behavior as parents will induce positive caring attitudes in terms of providing the best environmental conditions to promote infant development. Once again, the specific situation of the breastfeeding act provides unique and powerful evidence to support mothering behavior.

The Breastfeeding Act as a Unique Scenario for Bonding and Attachment

The act of breastfeeding provides a unique interaction where the mother will satisfy not only the nutritional but also the emotional needs of the infant. This scenario provides strong stimuli to favor the development of positive mother-infant interactions that have early and long-term consequences. During the act, mother and infant are interconnected not only physically but also sensorially, physiologically and psychologically. The complexity and richness of breastfeeding cannot be replicated to the same extent during artificial feeding.

Human offspring are highly sensitive to tactile stimuli, and touching is inevitable during the act of breastfeeding. The act requires that mother and infant be in close proximity and thus receive pleasure from each other. The infant will feel warmth and may be able to recognize the mother's heart beat, both stimuli that were provided constantly by the mother during intrauterine life. The provision of these stimuli may favor a more gradual transition to the abrupt changes the infant experiences after leaving the uterine environment [7].

Olfactory stimuli are also unique to the breastfeeding act. Maternal odors, especially breast milk odor, can be recognized by the infant. Recent studies have demonstrated that infants are able to respond to their own mother's breast milk odor, from the first weeks of life, exhibiting orientation behavior towards its source from the second week of life. Whether these stimuli trigger additional hormonal or physiologic responses remains to be determined [8].

Visual contact between mother and infant during breastfeeding has been studied. Visual contact is a powerful interactive stimulus in humans. During breastfeeding, mother and infant gazes are locked to each other more often than during bottle feeding. This provides the mother with a good opportunity to check her infant closely and evaluate if reality matches her expectations and fantasies. It will help her to create a real image of her child and incorporate it into her consciousness [7]. Moreover, this interactive behavior is modified by the composition of the breast milk: low-fat breast milk is associated with a longer time at the breast per feed and longer face-to-face mother-infant gazing behavior. Hormonal responses in response to sucking will in turn increase fat yield as lactation progresses [9]. Since breastfed infants need to spend more time sucking and actively feeding than children fed by the easy delivery system provided by the bottle, they are more exposed to the maternal stimuli accompanying the act of feeding. Whether this longer exposure to physiologic, sensory and psychological stimulation has a positive influence on infant development remains to be determined. Despite this uncertainty, it is clear that the bottle-fed infant will not be fed in quite the same manner as one that is breastfed.

From the mother's perspective, the breastfeeding act also provides unique stimuli and responses that are not the same if she bottle feeds her infant. The act is associated with a variety of sensory, physiologic and psychologic stimuli. In response to nipple suction, visual and olfactory cues, the mother triggers hypothalamic responses which mediate oxytocin and prolactin release. These hormones are responsible for successful breast milk production and release; moreover, they have a potential effect on maternal behavior fostering 'mothering' responses. These hormonal changes will induce uterine contractions, increased skin blood flow and temperature which will offer a sense of pleasure similar to that of sexual activity. In fact, the emotional and physiologic responses associated with nipple suctioning are quite similar during breastfeeding and sexual foreplay. These phenomena may not be easily recognized by mothers since there is an apparent contradiction between maternal and sexual emotional behavior. Women with an adequately developed sexuality report experiencing pleasure during the breastfeeding act. If their sexuality is perturbed, the pleasure sensations during breastfeeding may be associated with feelings of guilt. This knowledge is important when counseling is provided to breastfeeding mothers [4, 7].

A successful breastfeeding experience has additional benefits for mothers and for their role in fostering infant development. It reaffirms their feminine role to themselves and to society. This almost universal concept has important implications for self-image and self-esteem. Most societies place a high value on women's ability to care for their infants. In fact, in some societies all other merits are judged after this first and essential condition has been met. This leads to a negative feeling on the part of women who are not able to breastfeed successfully and may interfere with their role in infant development. On the other hand, a woman who breastfeeds successfully will reaffirm the typical feminine characteristic of satisfaction through caring for others. She will see her contribution to her infant's growth and development as a prolongation of her fostering role during pregnancy and will obtain satisfaction from this [10].

In evaluating the unique responses of mothers and infants to the breastfeeding act, it is necessary to state that by providing an adequate orientation to mothers, some of the positive conditions may be reproducible during artificial feeding. In situations where mothers are not able to breastfeed, they can still favor close physical proximity, visual contact and tactile stimulation. Nevertheless, the physiologic effects on mothers and infants and the cascade of events that relate to this unique interaction cannot be assimilated to the act of bottle feeding. The interactive responses of the dyad are reduced. The risk of approaching bottle feeding as a mechanical act may further compromise the opportunity for using the feeding situation to foster development. This is particularly important, since in today's world the feeding situation is the main and often the only opportunity to establish intimate and prolonged contact between mother and infant.

Methodological Approaches to Assess the Relationship between Type of Feeding and Developmental Outcomes

A critical review of the published information on the effect of breastfeeding and developmental outcomes reveals important methodological problems. The first is the nature of the experimental design used to address the question and the second is the lack of standardization in methods used to assess infant development, which makes comparisons between studies difficult.

Most studies are retrospective and results are derived from secondary analysis of data collected for other purposes. The data on breastfeeding collected using this approach are fragmentary at best: only decision to feed and duration of breastfeeding are available in most studies. Maternal characteristics, weaning practices, and difficulties encountered by the breastfed group are usually lacking. Important confounding variables such as quality of the home environment, mother-infant interaction, maternal depression and attitudes towards infant development are also scant or assessed through unreliable methods. The measured intervening variables usually correspond to the primary objective of the study, where feeding mode itself is a confounder. This translates into experimental designs which are inadequate or insufficient to determine if breastfeeding is the main reason for the putative changes described. If we are to have definite answers to this crucial question, we will need prospective longitudinal studies where the confounding variables are carefully recorded as early as possible. The relevant prenatal maternal variables, as well as the environmental conditions that may affect infant development, need to be obtained. The occurrence of other factors which may affect mental development must be prospectively recorded, and not dependent on maternal memory or her subjective perception of the event.

The variability in methods to assess development makes the comparison between studies extremely difficult. Although infant psychomotor development is usually evaluated using the Bayley Scale of Infant Development (BSID) or at least the mother scale, studies with a follow-up beyond 2 years of age have used the McCarthy, WISC, Peabody, Stanford-Binet and various other cognitive measurements.

Cognitive outcome for the 0 to 2-year period are predominantly based on psychomotor development scales which have poor predictive value for infants that score within the normal range. New methods are available that permit the evaluation of specific aspects of development, such as visual cognitive maturation, which have better predictive value.

Methods to Evaluate Infant Development (0–2 Years)

Bayley Scale of Infant Development. There is no doubt that the BSID continues to be the most widely used method. It provides mental and motor developmental indices and is supported by a strong normative framework. The new 1993 revised edition is extended to 36 months of age. It overcomes previous deficiencies in relation to the assessment of visual perceptive skills and it has an improved behavioral record form [11].

Novelty Preference. Fagan and colleagues [12–14] have developed a highly standardized methodology based on the assumption that if an infant prefers one stimulus over another presented simultaneously, he or she can discriminate between them. If the infant prefers the novel picture, he or she must also be able to recognize and remember what was previously seen. A child's preference for a novel or familiar stimulus depends on the habituation process. Before habituation has occurred, infants presented with a familiar and a novel stimulus simultaneously will demonstrate a preference for the familiar one. After habituation, novelty preference will prevail. This experimental paradigm has been used to characterize an infant's ability to discriminate patterns, faces, colors, shapes, movements, orientations and logical arrangements. An important advantage of this and other methods that evaluate visual cognitive behavior is that they are not dependent on major motor capacities, making it possible to study visual discrimination and recognition memory during the first 6 months of life. Data published by Fagan et al. [15] suggest that this approach to evaluating high-risk and normal infants achieves higher correlations than motor sensory testing for predicting normal or abnormal mental development indices at 3 years of age. The sensitivity of their test for predicting mental developmental indices <70 was 100%, while its specificity was over 90%. The long-term validation of visual perceptual cognitive behavior is still in progress. On the basis of preliminary evidence, the test appears to be a promising new tool for evaluating the effect of early diet on brain and visual development.

Mastery Motivation. Another interesting and new approach to cognitive measurement is the evaluation of motivational aspects of behavior. Morgan et al. [16] have developed a standardized methodology to assess mastery motivation in infants aged 9 to 36 months. Mastery motivation is defined as the psychological force that stimulates an individual to attempt, in a focused and persistent manner, to solve a problem or master a skill or task which is at least moderately challenging. During the test situation, the child is invited to solve a preestablished age-adjusted task. Persistence at task and affect are coded every 15 s, for a total of

4 min. Three kinds of tasks, with different levels of difficulty, are available: puzzles, shape sorters and cause/effect toys. This test is more related to temperament and motivation than to actual cognitive skills. It may serve to explain part of the variance in cognitive tests that are not related to intellectual capacity.

Methods to Evaluate Child Development (> 2 Years)

Developmental studies of preschool children are usually based on intellectual-capacity scales; in some, school performance is also added. Cognitive development is a complex issue; its evaluation requires the assessment of a range of functions. The battery to be used should include tests of intellectual capacity, language skills, visual-motor coordination, and psychoeducative abilities. At the same time, the evaluation should be complemented with measurements of motor skills and of social and emotional development.

To assess intellectual function, the Stanford-Binet Intelligence Scale (Form L-M, 3rd revision [17]) has the advantage of providing an evolutionary design; this is especially helpful for longitudinal studies where repeated measurements are taken. This test provides an index of general intelligence (IQ) with an average of 100 and a SD of 16. It also allows a factorial analysis that evaluates six factors: general comprehension, judgement and reasoning, arithmetic reasoning, visual motor ability, vocabulary verbal fluency, and memory and concentration [17, 18].

For the assessment of linguistic abilities, the main choice is the Illinois Psycholinguistic Abilities Test [19]. This language test evaluates cognitive functions that intervene in the communication process. It comprises 10 subtests, mainly oriented to the assessment of receptive and expressive processes. Each subtest provides an equivalent age and standard score. The global analysis of the test provides a language measure that can be expressed as a psycholinguistic quotient and/or a standard score average.

For psychoeducational abilities, the Short Preschool Scale of the Woodcock test [20] seems to be a good alternative. It includes quantitative concepts, picture vocabulary, spatial relations and visual-auditory learning. This test provides an age score of broad cognitive ability. It also allows the estimation of percentile ranks, standard scores, normal-curve equivalents and relative performance indices.

The Bruininks-Oseretsky Test of Motor Proficiency [21] provides a comprehensive index of general motor proficiency as well as separate measures of both gross and fine motor skills. Normative data include standard scores for age groups, percentile ranks and stanines. Age equivalence is also provided.

For the assessment of visual-motor integration, the VMI test [22] provides a comprehensive evaluation. Raw scores obtained after copying 24 geometrical forms are compared with normative data, allowing an estimation of age equivalence, percentiles and standard score.

Assessment of Confounding Variables

All the studies we have reviewed have insufficient assessment of confounding variables. The studies have considered gross sociodemographic variables but not the type of variables that are potentially responsible for the subtle effects expected from feeding modality. It is crucial to include variables that evaluate maternal-infant interaction and the quality of the environment in terms of the stimuli necessary for normal development. Assessing this question with our present knowledge, it is clear that the determinants of breastfeeding are also involved in defining cognitive development. Thus the separation of these covariates is virtually impossible based on their developmental effects. Social class, the parents' educational level, birth order and parental interest in the child's education are all related to the decision whether or not to breastfeed. These same factors have demonstrable effects on cognitive development, as demonstrated by multiple studies. Thus, unless they are considered in the evaluation of cognitive outcome, the dilemma of whether the effect observed is due to who feeds the child and how it is fed, rather than to what is fed, will not be resolved.

Moreover, there are many other confounding variables that probably also affect mental development that have not been measured. These relate to the mother, the infant, their interaction and the role of the environment. The lack of inclusion of these variables in the analysis may lead to an erroneous interpretation of results. The putative benefits of breastfeeding may in fact be accounted for by the presence of other covariates which affect development.

By their very nature, these variables must be measured by specially trained professionals using standardized and sensitive instruments in order to obtain reliable results. We will briefly review the main confounding variables and how they can be measured if we are to resolve the proposed dilemma.

Mother-Infant Interaction. There is currently great controversy concerning the best way to evaluate the mother-infant relationship. The instruments available at the moment have been developed mainly as clinical tools. That is, they have been designed to orient the clinical psychologist in strengthening the mother-infant relationship [23]. A potentially better approach, which is now being used by us and others, is the videorecording of a standardized play situation which is then coded and scored in a standard manner by an independent assessor who is not familiar with the dyad being evaluated. In our studies, we record the interaction for 20 min in each of three experimental situations. In the first, we ask the mother to accompany her child while she or he plays with a new set of toys. Then we request that she teaches the child a new task and finally we ask her to play with the child as she would do customarily. These three situations provide a sample of maternal-infant interactions which is closer to real life than the instruments that have been used traditionally. A comparison of the two approaches illustrates a theme common to all behavioral testing: instruments appropriate for clinical testing are usually not appropriate to evaluate cases that are within the normal range but may still be quantitatively and qualitatively different from each other.

Home Inventory. Evaluation of stimulation in the home environment has become a must in developmental studies. The inventory developed by Caldwell [24] is the best known alternative. There are three versions of the Home Inventory, one for infants, one for toddlers and another for school-children. The infant version comprises 45 items including emotional and verbal responsiveness of the mother, avoidance of restriction and punishment, organization of the physical and temporal environment, provision of appropriate play materials, maternal involvement with the child and opportunities for variety in daily stimulation. Assessment takes place at home in the presence of the mother and child together; some of the items are observed directly and others are reported by the mother or caretaker. As a consequence of our experience in applying the Home Inventory, we have added additional items to measure other relevant aspects of child development that were particularly relevant to malnourished infants. We established items to assess paternal participation, the child's daily routine, sibling interactions and social networks. We are analyzing these components separately and believe they will add further strength to the Home Inventory [24, 25].

Maternal Depression. There are several choices to assess mothers' depression. We have favored simplicity, particularly when studying women of low socioeconomic level since they often have difficulty understanding abstract questions. We consider the self-reporting Center for Epidemiological Studies Depression Scale (CES-D) [26] short and easy to use. It provides an index of cognitive, affective and behavioral depressive features and the frequency of occurrence during the preceding week. The assessment of this variable recognizes the mother's crucial role in early development. This tool permits the inclusion of maternal depression in practice.

Maternal Intelligence. Since the mother is the key modulator of the child's cognitive environment we consider that the evaluation of maternal intelligence through the Wechsler Adult Intelligence Scale (WAIS) to be highly relevant. On the basis of our prior experience, we have selected four subtests: social comprehension, similarities, block design and picture completion. Our studies have demonstrated that this shorter form is timesaving and highly correlated with the full-scale IQ when comparing intellectual performances of groups [27].

Effect of Feeding Type on Developmental Outcomes

Effects of Human Milk Feeding in Term Infants
The first studies proposing that breast milk feeding might exert a beneficial effect on infant psychomotor development were published during the 1920s. Since then, several studies have reported small but significant advantages in

Table 1. Breastfeeding and cognitive outcomes in seven studies

Authors	Subjects and design	Functions assessed	Results	Variables associated with breastfeeding
Rodgers, 1978 [28]	1,133 BF 11,291 FF 8- and 15-year-olds retrospective	picture intelligence and word reading at 8 years; nonverbal ability and math attainment 15 years	FF: 0.5–1.7 points lower differences linked to: feeding; social class; home material condition; parent education; family size, and birth rank	BF associated with: socioeconomic status; parents' education; parents' interest in child education, and child's birth rank
Taylor and Wadsworth, 1984 [33]	13,135 5-year-olds retrospective	English picture vocabulary (EPV) copying design child behavior score maternal malaise	differ in EPVT copying and design and in child behavior differences lost after adjusting for height and head circumference	BF linked to social class, low birth weight, and maternal score
Ferguson et al., 1982 [30]	1,037 3-year-olds 3,991 5-year-olds 954 7-year-olds retrospective	Peabody picture vocabulary test at 3 years Stanford and Reynell at 5 years WISC and Illinois at 7 years	differences in IQ and language depending on length of BF (none, <4 months, >4 months) after adjustment, BF (group) 2 points higher in each category	
Morrow-Tlucak et al., 1988 [31]	157 FF 39 BF < 4 months 23 BF > 4 months retrospective	BSID-MDI	no differences at 6 months; at 12 and 24 months, MDI 2.1 higher for each BF category	groups differ in following maternal variables: IQ; authoritarian behavior; smoking and age
Bauer et al., 1991 [31]	50 3-year-olds prospective	McCarthy Scales of children's ability	BF length linked to memory, verbal, quantitative and general cognitive abilities	
Jacobson and Jacobson, 1992 [32]	323 4-year-olds retrospective	McCarthy Scales of children's abilities	BF had greater verbal abilities differences lost adjusting for mother's IQ and parental abilities	BF associated with: maternal education; social class; material IQ, and parental abilities
Rogan and Gladen 1993 [29]	850 assessed at 6, 12, 18 and 24 months, 3, 4 and 5 years retrospective	BSID McCarthy Scales of children's abilities	differences in MDI at 24 months and in McCarthy at 3 and 4 years	

BF = Breastfed; FF = formula fed; MDI = mental development indices.

developmental indices for breastfed infants, although no study has categorically proven cause and effect. This is not surprising, given the acknowledged limitations in the experimental approaches in this area of research. Over the past decade, interest in this question has revived. Most of the recent studies demonstrate better developmental scores during infancy and preschool years, with marginal advantages during school age especially in the cognitive and language areas of development.

Early effects have been detected mainly in infants' psychomotor performance. The advantage for the breastfed child in language-related and visual cognitive skills may reach 6–10 points compared to bottle-fed controls. After correction for confounding variables, the differences decrease to 3–5. In children older than 2 years of age, the main differences are found in the language and visual-motor coordination areas. The differences favoring the breast-fed child range from 4 to 10 points. The fact that the differences relate predominantly to mental performance and language skills supports the hypothesis that the principal effects are mediated by maternal-infant interactions rather than by a generalized benefit in central nervous system development, although this does not exclude the possibility that the improved interaction may be mediated by a biological effect at a critical time in the development of, for example, visual or auditory sensory systems. The evidence to date indicates that the effects of early stimulation in vulnerable infants are manifested by improved cognitive and language skills, while the effects of specific nutritional interventions are related to improved motor performance. The characteristics of studies in full-term infants and the main results are summarized in table 1. Most are retrospective in nature [28–34].

In the Rodgers study [28], one of the few long-term follow-up studies of mental development comparing breastfed with formula-fed children, where family, social and economic variables were controlled, early breastfeeding was associated with better picture intelligence at 8 years of age, and better scores in mathematics, nonverbal ability and sentence completion at 15 years of age. The study cohorts were selected from all live births occurring during 1 week in 1946 in Great Britain. Based on confirmed records of early diet, 1,133 individuals were considered to have been entirely bottle fed and 1,291 individuals were never bottle fed. The functional benefits were proportional to the duration of breastfeeding and remained statistically significant after controlling for social, cultural and demographic variables by multivariate analysis. These observations could not be extrapolated to today's formula-fed infants because during the observation period children were fed unmodified, diluted cow's milk with the sole addition of sugar. Protein may have been excessive and essential fatty acids (EFAs) may have been inadequate. However, a recent study by Rogan and Gladen [29] of a cohort of 855 newborns, enrolled from 1978 to 1982 and followed through school age, for whom present-day formula foods were used and modern methodology applied to assess cognitive development, confirmed the results of the earlier British study. Breastfed infants had significantly higher scores in the BSID at 2 years and in the McCarthy Scale at 3 and 4 years. In addition, slightly higher English grades on report cards were found in the breastfed children after adjusting for the relevant confounding variables.

Present evaluations of infant development cannot ignore the evaluation of confounding variables which affect developmental outcome. The interactive effects of biologic-genetic, family environment, and sociocultural environmental factors determine a large proportion of the variability observed in developmental indices.

Most of the studies summarized in table 1 controlled for biodemographic factors such as birth weight, birth rank, family socioeconomic level, parental education and maternal age. These variables have demonstrable effects on cognitive development, but they explain only a small proportion of the variability in the assessment indices. More important than these are the home conditions, maternal psychologic attributes, child-rearing practices and mother-infant and family interactions. Few of the studies detailed in table 1 controlled for these important factors: four controlled for parental child-rearing capacities [28, 30–32], three controlled for maternal IQ [30–32], while only one controlled for home environment stimulation [31], and again only one for maternal mental health [33]. The best in terms of controlling for confounding variables were the Ferguson et al. [30] and Morrow-Tlucak et al. [31] studies.

The other aspect which makes the interpretation of results complex if not virtually impossible is the covariation associated with breastfeeding. Breastfed infants belong to families with higher socioeconomic status, parents with more education, mothers with a higher IQ and fathers with higher educational achievement and more interest in child development; moreover, these mothers and fathers also exhibit better parenting skills. Most, if not all, of these variables are clearly and probably causally related to improved mental development across feeding modalities and usually underlie the observed benefits derived from breastfeeding in term infants. The dilemma of establishing a causal relationship in such a confounded situation cannot be solved by statistical processing since it is never possible to really fully correct for covariates. The prospects for performing case-control studies using a sufficiently large sample size should be explored, since it is ethically impossible to do prospective randomized controlled trials of breastfeeding.

Another key variable that has not yet been considered is the importance of the developmental level of the maternal ego and the role of breastfeeding in enhancing her ego. Studies of maternal factors which affect the decision to breastfeed have established that a strong maternal ego increases the likelihood of successful breastfeeding. Mothers with mature egos demonstrate greater empathy, nurturing capacity, responsibility and a greater capacity to enjoy caring for the infant. All these characteristics indicate a capacity for centering on the child rather than on the self. This is clearly a prerequisite for successful breastfeeding. A mature ego is usually less egoistic and less authoritarian; these traits are inversely related to the success of breastfeeding. Again, the same characteristics linked to a mature ego have a positive effect on the mother-infant interaction and on infant care practices. Such mothers show a greater sensitivity to infant's needs, more flexibility in rearing practices and are more able to generate a stimulating atmosphere for the infant, all of which will favorably influence the infant's cognitive development. In the study by Jacobson et al. [35], the effect of a better home stimulation environment, assessed using the Home Inventory, was found to be related to the prevalence of breastfeeding. Mothers who breastfeed provide more favorable play situations, more and a greater variety of opportunities for stimulation and are generally more involved in infant care and stimulus. These findings are especially revealing, since the study was conducted with mothers of low socioeconomic level. These differences in the home stimulation environment are probably related to the previously described ego characteristics, although this was not specifically addressed in the study.

Most studies presented in table 1 used multiple regression methods to control for confounding covariates. They are consistent in showing that initial differences in socioeconomic level and maternal education are significant determinants of development across feeding methods. Correcting for these variables diminishes the differences but the effects remain significant. In only two studies were the differences lost after developmental scores were adjusted. In one study differences were lost after correcting for height and head circumference at 5 years, but this correction may not be fully valid since it cannot be excluded that these growth measures may in fact be related to breastfeeding. In the second study, difference in the McCarthy test disappeared after adjusting for maternal IQ and parenting capacities. This was probably a valid correction and stresses the need to include maternal variables as potential confounders [32, 33]. The magnitude of the advantages for breastfed infants after adjustment is in the range of 3–5 points. This may appear small by most criteria, but when one considers that they are in the range of the observed benefits after correcting a mild lead burden or iron deficiency anemia, the social consequences become clearer. Since feeding mode is a variable that affects all infants, it is expected that improving a few percent points in mental developmental indices may provide a great benefit to a population group.

Effects of Human Milk Feeding in Preterm Infants

The beneficial effects of human milk feeding on mental development appear to be more pronounced in preterm very-low-birth-weight (VLBW) infants. Although the initial collaborative studies from England demonstrated a disadvantage for banked human milk, in 1988, Morley et al. [36] described the favorable effect of human milk fed to 771 low-birth-weight (LBW) infants at 18 months of age. Infants from mothers who decided to provide human milk for them had a developmental index 8 points higher than those fed formula. This difference decreased to 4.3 points after controlling for perinatal and demographic confounding variables. In a recent study, Lucas et al. [37] reported that preterm infants with a birth weight < 1,850 g whose mothers provided human milk 72 h postpartum had a 10-point advantage in cognitive skills assessed by an abbreviated WISC test at 8 years of age. Considering that the infants were tube fed either their mother's milk or formula, the difference suggests that there may be biological factors in human milk responsible for the developmental differences. Furthermore, infants from a group of mothers who decided to provide human milk but were unsuccessful at establishing a milk supply had scores similar to those fed formula. After adjusting for confounding variables such as maternal socioeconomic status, parental educational level, sex and ventilator treatment, the differences decreased to 8 points but remained statistically significant. Unfortunately, the studies failed to control for other maternal variables, home environment and stimuli provided. Thus, the results are suggestive but clearly not conclusive. Moreover, the decision to provide human milk may indirectly reveal the maternal commitment to the infant's well-being and development beyond the immediate neonatal period, thus confounding the interpretation of the results. The fact that the specific difference in infant groups was solely the provision of human milk for the initial 28 days of life suggests that there may be factors in human milk which could play a critical role in brain development and maturation. Because infants were not placed at the breast but were tube fed, the potential effect of the breastfeeding act was controlled.

The authors indicated that among the remaining differences between human milk and modern preterm formula, the presence of long-chain ω3 fatty acids in human milk should be considered as a possible factor in determining the effect. This topic is discussed below.

Essential Fatty Acids as Determinants of Retinal and Brain Development

Essential Fatty Acids

Pediatricians have considered PUFA to be part of the lipid supply necessary for energy required for growth, cellular metabolism and muscle activity. The fact that some PUFAs are also EFAs, and that they serve as dietary precursors for eicosanoid and docosanoid formation, has provided greater significance to their study. During this decade, attention has focused on the role of ω3 PUFAs in the prevention of cardiovascular disease and as immunomodulators. The potential

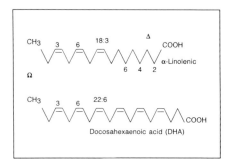

Fig. 1. Structure and nomenclature used to describe the main ω3 fatty acids.
α-Linolenic acid: *cis,cis,cis*-Δ^9,Δ^{12},Δ^{15}-octadecatrienoic acid, an 18-carbon chain with
cis-configuration double bonds positioned at C_9, C_{12} and C_{15} as numbered from the
carbonyl (Δ) terminus, or 18:3ω3 when numbered from the methyl(ω) terminus. DHA:
all *cis*-4,7,10,13,16,19-docosahexaenoic acid, or 22:6ω3.

role of PUFAs as structural components of membrane phospholipids necessary
for normal eye and brain development is increasingly being recognized [38–41].
ω3 fatty acids, especially docosahexaenoic acid (DHA C22:6ω3), present in
human milk have been shown to be necessary for retinal and brain development in
primates and humans [38–41]. Rod photoreceptor function and the maturation of
visual acuity of human LBW infants are dependent on the supply of these essential
nutrients [38–41]. The visual function of full-term infants fed human milk is
enhanced for up to 3 years, supporting the concept of long-term benefits of human
milk feeding on mental development [42]. LBW infants fed human milk or
formula by tube served to demonstrate the benefits of human milk on IQ at 8 years
of age [37].

The essentiality of ω6 and ω3 fatty acids for humans is best explained by the
inability of animal tissues to introduce double bonds in positions prior to carbon 9
counting from the ω terminus [40, 43]. Figure 1 presents the structure and
nomenclature used to describe the ω3 fatty acids. Burr and Burr [43] were the first
to characterize the essential nature of PUFAs and proposed that three fatty acids
be considered essential: (a) linoleic acid (LA; 9,12-octadecadienoic acid), 18:2ω6
because of a double bond at carbon 6 counting from the methyl end (ω carbon) of
the molecule; (b) arachidonic acid (AA; 5,8,11,14-eicosatetraenoic acid), 20:4ω6,
because it possesses 20 carbon atoms and four double bonds one of which is at the
ω6 position, and (c) α-linolenic acid (LNA; 9,12,15-octadecatrienoic acid),
18:3ω3, having 18 carbon atoms and three double bonds, one of which is at
carbon ω3, the equivalent of Δ^{15} following the accepted IUPAC nomenclature
starting from the carboxyl terminal.

The biochemical pathways for ω6 and ω3 fatty acid desaturation are only
present in chloroplasts. Thus, only higher plants, algae, and some fungi are ca-

pable of forming these EFAs. Terrestrial or marine plants are the primary source of EFAs in the food chain; fish and other marine animals are able to elongate and desaturate the parent EFAs forming the long-chain PUFAs.

Hansen et al. [44] firmly established that LA is essential for normal infant nutrition, in a clinical and biochemical study of 428 infants fed cow's-milk-based formulations with different types of fat. The daily LA intake of study infants ranged from 10 mg/kg when fed a fully skimmed milk preparation to 800 mg/kg when a corn/coconut-oil-based preparation was fed. They observed dryness, skin desquamation and thickening and growth faltering as frequent manifestations of LA deficiency in young infants.

EFA deficiency is associated with specific findings in infants and young animals [45–48]. The syndrome has usually been related to LA deficiency but, in most models, a LNA deficiency coexists. ω3 deficiency induces subtler changes including skin changes unresponsive to LA supplementation, abnormal visual function and peripheral neuropathy [49]. The nervous system manifestations are probably a response to an insufficiency of the specific metabolic derivative of LNA, namely DHA. Indeed, the high concentrations of DHA in the cerebral cortex and retina would suggest its participation in neural and visual function [50, 51]. DHA reaches levels of up to 50% of the total fatty acids in the phospholipids of these tissues, yet a specific role for DHA in the physiological or biochemical function of neural tissues remains to be fully defined. Recent studies have attempted to characterize the biochemical and functional effects of dietary ω3 deficiency. Animal models have been developed using purified LA as the only dietary fat or using safflower or sunflower oil in which the ω6/ω3 fatty acid ratio (approximately 250:1) is very high [52–54]. Similar high ratios are found in powdered infant formulas currently in use in some parts of the world [55, 56].

Studies in Term Infants

The plasma DHA concentrations of full-term infants fed formula are lower than those of breast-fed infants. This suggests that present formulas either provide insufficient LNA, or chain elongation-desaturation enzymes are not sufficiently active during early life to support tissue accretion of DHA. Full-term infants may also be dependent on a dietary DHA source for optimal functional maturation of the retina and visual cortex [42]. Furthermore, studies in infants born at term, dying from sudden infant death, revealed that brain composition is affected by human milk feeding in terms of a higher DHA content in the brain cortex of breastfed infants relative to those receiving cow's-milk-based formulas [57]. However, no controlled trials of DHA supplementation of full-term infants are available to date.

We studied the effects of postnatal age on the maturation of visual acuity in full-term infants fed either human milk or cow's milk formula containing 12–18%

LA and 0.5–1.0% LNA [42]. These studies indicated that conceptional age, not postnatal age, determined visual maturation. Furthermore, indices of visual acuity were maturer in 4-month-old, exclusively breastfed infants in comparison to formula-fed infants. At 4 months of age, visual acuity measured using evoked potentials and expressed in Snellen equivalents, were a mean of 20/85 for the formula-fed and 20/65 for the human-milk-fed infants. Behaviorally assessed acuities gave means of 20/130 and 20/110, respectively ($p < 0.01$) [42].

We also completed a 3-year follow-up of healthy, full-term infants who participated in a controlled study of the effects of dietary fat on lipoprotein metabolism [42]. The cohorts were either breastfed from birth to at least 4 months or were fed formula for 12 months containing ample LA and 0.5% of the total energy as LNA. The breastfed group was weaned to an oleic-acid-predominant formula and received egg yolk as a source of cholesterol. The egg yolk also provided 80–100 mg DHA per day. The breastfed group maintained higher plasma and red blood cell (RBC) membrane phospholipid DHA concentrations throughout the first year of life. At 3 years of age, stereo acuity was measured by operant preferential looking (OPL) techniques. Mean ($\pm$SD) OPL stereo acuity was 42.1 ± 5 for the breastfed and 92.8 ± 86.1 seconds of arc for the formula-fed groups ($p < 0.05$). Since values varied greatly in the formula group, the data were analyzed nonparametrically. This analysis showed that 92% of the breastfed group had an OPL stereo acuity of <40 seconds of arc (considered fully mature), whereas only 35% of the formula-fed group met that criterion ($p < 0.001$). Monocular acuities assessed by OPL were similar in the two groups (mean Snellen equivalents of 20/28 and 20/32, respectively, $p < 0.16$). Visual recognition was assessed by the child's ability to match three letters. In the breastfed group, the mean ($\pm$SD) score was 2.71 ± 0.8, while in the formula-fed group, it was 1.82 ± 1.5 ($p < 0.04$). 93% of the breastfed group had a perfect score (3 of 3) while only 61% in the formula-fed group did so ($p < 0.001$). It is notable that these differences could be found at 3 years of age in this select group by observers who were blinded to the dietary group assignment [42].

Studies in Preterm Infants

The purpose of our past studies in Dallas (USA) and present research carried out in Santiago (Chile) is to assess the essentiality of ω3 fatty acids in humans. We believed that VLBW infants were particularly vulnerable to deficiency given the virtual absence of adipose tissue stores at birth, the possible immaturity of the fatty acid elongation and desaturation pathways and the inadequate LNA and DHA intake provided by formula. Over the past decade, we and others have conducted studies to evaluate the effect of ω3 fatty acids in VLBW infants examining the effects of LNA, or LNA plus DHA supplementation on plasma and tissue lipid composition, retinal function, on maturation of the cortex and on measures of infant growth and development [58–64].

In a series of studies, we characterized the biochemical and functional effects of dietary ω3 fatty acid deficiency in preterm infants. In 1986, low LNA was found in powdered infant formulas in most countries. Now, virtually all infant formula is supplemented with LNA and some manufacturers in Europe have added DHA, or DHA plus AA, to preterm formula. In addition, we and others have demonstrated that long-chain PUFAs, such as DHA, provide a specific structural environment within the phospholipid bilayer influencing important membrane functions such as ion or solute transport, receptor activity and enzyme action [65–67].

The assessment of visual function was selected as a sensitive index of the subtle actions of ω3 fatty acids in human infants, given the demonstrated effects on visual development in vertebrates. Replacement of ω3 by ω6 or ω9 fatty acids in the retina and cortex impairs visual and brain functions [52–54, 68]. A significant correlation between brain and RBC membrane phospholipid DHA composition during deficiency has also been established in animal studies [69, 70].

We studied 83 newborns with body weights of 1,000–1,500 g, 28–32 weeks gestation at birth, receiving enteral feedings and free of major neonatal morbidity by day 10 of life. Ten infants receiving human milk served as controls for the study, and we also evaluated 12 infants who had remained in utero and tested them at the equivalent conceptional age relative to the study infants. The group fed human milk was supplemented with human milk fortifier (Enfamil, Mead-Johnson) and received ≥75% of their intake as their own mother's milk up to 36 weeks postconceptional age (usual discharge age). If mothers were unable to fully provide human milk, the soy/marine oil formula was used to supplement their feeding. The remaining 73 infants were randomly assigned to one of three formula groups varying in EFAs. Experimental formula feedings began at day 10 of life and continued until 57 weeks postconceptional age (i.e., equivalent to 4 months postterm) [58–60].

The EFA composition of the human milk and study formulas has been published [59]. Briefly, the corn-oil-based formula corresponded to commercial powdered premature formula and contained 24% LA, was extremely low in LNA (0.5%) and had no long-chain PUFAs; the soy-oil-based formula was equivalent to current ready-to-feed premature formula used in the USA; the third study formula was also soy oil based, with 1.4% LNA, but was enriched with marine oil to give an ω3 long-chain PUFA content closer to that found in human milk [8]. At the time we started our study, ω3 long-chain PUFAs could only be obtained from marine oils [58–60].

Full-field electroretinographic (ERG) responses to short- and long-wavelength stimuli over an extensive range of retinal illuminances were evaluated at 36 and 57 weeks postconception. The ERG response to short-wave stimuli was used to isolate rod photoreceptor function at the two test ages. Naka-Rushton plots were computed from rod responses to graded illuminances; threshold, maximum amplitude and semisaturation constants were calculated [58, 60].

Visual-acuity development measured by pattern reversal visual evoked potentials (VEPs) was evaluated at the 36- and 57-week follow-up. Behavioral measurements of visual acuity [i.e., forced-choice preferential looking (FPL) visual-acuity responses] could be reliably tested only at the 57-week follow-up [61, 71, 72]. The VEP and FPL visual-acuity tests are used to evaluate the neural integrity of the pathway from retina to primary visual cortex and the ability of the infant to see and produce a motor response when visually stimulated [71, 72].

The composition of RBC lipids for all diet groups was similar on entry into the study [59]. At the 36- and 57-week postconception follow-up, marked differences in DHA content were evident in the RBC lipids among the diet groups. Infants fed the corn oil formula deficient in ω3 fatty acids had significantly lower DHA levels than the other formula groups. The marine oil group presented DHA concentrations elevated above all other groups. The relative content of docosapentaenoic acid (DPA) found at 36 and 57 weeks in RBC lipids was inversely related to the DHA content. DPA was significantly elevated in the corn oil group in comparison to the human milk and soy/marine groups. By 57 weeks, the differences in the end products of ω3 and ω6 fatty acid metabolism, namely DHA and DPA, were greatly accentuated compared to 36-week values. The group fed corn oil was significantly different in both DHA and DPA content compared to all other diet groups. RBC lipids of the human milk and marine oil groups were statistically different in DHA composition but showed similar patterns for both total ω3 and ω6 long-chain PUFAs. The soy oil group had intermediate DHA and DPA values relative to corn and soy/marine oil groups but differed significantly from the DHA-supplemented infants [59, 73].

Retinal-function responses demonstrated significantly higher threshold values from rod photoreceptors in the ω3-deficient group (corn oil fed) at the 36-week follow-up relative to groups receiving ω3 fatty acids [58, 60]. The light intensity required to elicit a 2-μV response (threshold) was greater when plasma DHA levels were low, indicating that the sensitivity and maturity of the rod photoreceptors in ω3 fatty acids deficient infants was reduced significantly compared to infants fed soy/marine formula or human milk. A group of 10 infants born at 35 weeks had their visual maturity tested a few (3–5) days after birth and served as a normal standard for comparison. This group of infants receiving EFAs transplacentally was nearly identical in all rod ERG functional indices to premature infants of the equivalent conceptional age fed human or marine oil. The group fed soy oil had higher threshold values than the group fed soy/marine oil (p < 0.06). These results are summarized in table 2. Cone photoreceptor function was not significantly affected by diet, although the trends were similar to the results obtained from rods. At the 57-week follow-up, retinal rods and cones showed no diet-induced differences in a or b wave parameters. At 57 weeks the infants fed corn oil had consistently longer implicit times in light-adapted

Table 2. Rod ERG (b wave) parameters of preterm infants at 36 weeks postconception according to early diet [adapted from ref. 60]

	Human milk	Corn oil	Soy oil	Soy/marine
log threshold	0.41	1.08***	0.71*	0.41
log V_{max}	1.20	1.05**	1.08	1.22
log k	1.25	1.73**	1.39	1.24

Threshold is measured in scotopic troland-seconds, a measure of illuminance. The threshold corresponds to the light required to elicit a 2-μV response. k corresponds to the illuminance required for half V_{max}, where V_{max} is the highest b wave response in microvolts. Normal preterm infants born 36 weeks postconception have an average log threshold of 0.27 scotopic troland-seconds, log V_{max} of 1.24 μV and log k of 1.12 scotopic troland-seconds. These values are not different from those in the groups fed human milk and soy/marine oil. *p < 0.06, **p < 0.005, ***p < 0.001: values differ significantly from human milk and soy/marine groups.

oscillatory potentials. The infants fed soy oil had some peaks differing significantly from those fed human milk, while the soy/marine and human milk groups were similar. Oscillatory potentials are generated in the inner retina [60].

At the 57-week follow-up, the groups fed soy/marine and human milk had lower log minimal angle of resolution, (MAR) values, that is better acuity, than infants fed formulas devoid of DHA using either VEP (electrophysiologic response) or FPL (behavioral) methods. A log MAR value of 0 corresponds to 20/20 Snellen equivalents while a value of 1 corresponds to 20/200. The infants fed human milk and soy/marine oil had the highest DHA/DPA ratios and also the best visual acuity (lowest log MAR). The group that received soy oil as a source of ω3 fatty acids had poorer VEP acuity at the 57-week follow-up relative to the group fed soy/marine oil, indicating that despite ample LNA, visual function was less mature than in the DHA-supplemented infants. A group of healthy full-term breastfed infants matched by conceptional age were used as controls. The groups fed human milk or soy/marine oil were virtually identical to the controls, while the corn and soy groups had poorer visual acuities [61, 72]. These results are summarized in table 3.

These studies provide clear evidence that dietary ω3 fatty acid deficiency affects the eye and brain function of preterm infants as measured by ERG responses, cortical VEPs and behavioral testing of visual acuity. Preterm infants require DHA in their diet because they are unable to form it in sufficient quantity from LNA provided by soy-oil-based formula products. Changes in ω3 and ω6 fatty acid intake resulted in discernible differences in the fatty acid composition of plasma and RBC membrane lipids.

Table 3. Visual acuity of preterm infants at 57 weeks postconception according to early diet [adapted from ref. 61]

	Human milk	Corn oil	Soy oil	Soy/marine
VEP	20/65	20/95*	20/85*	20/50
FPL	20/115	20/165*	20/140	20/125

Values significantly different from human milk and soy/marine groups at p < 0.05.
Normal full-term infants at 57 weeks postconception have an average 20/70 for VEP and 20/115 for FPL. * = Visual-acuity results are provided as Snellen equivalents, where 20/20 is average for normally sighted adults.

Despite efforts by many investigators, the American Academy of Pediatrics in the USA has still not acknowledged the need for and the essentiality of ω3 fatty acids; even today, low LNA formulas are still in use in some parts of the world [56]. On the other hand, the European Society for Pediatric Gastroenterology and Nutrition has recommended not only that LNA be present but has stated that it would be desirable that DHA and AA also be added to formulas destined for preterm infants [74].

We propose that the supply of essential PUFAs provided by human milk during early life influences the maturation of the retina and the brain. Our studies summarized here support an essential role for ω3 long-chain PUFAs for optimal visual development of preterm and possibly term infants.

Conclusions

The results comparing breastfeeding with bottle feeding presented here should be interpreted with caution considering the discussion of assessment methods and confounding variables. Undoubtedly, the effects of breast milk are complex, encompassing not only nutritional and psychological components, but also their interaction with biological development and emotional factors. We have discussed the effect of breastfeeding on mother and infant, but it is also possible that a given phenomenon may generate a cascade of physiologic or psychologic responses which may affect the dyad in a critical way. At present it is difficult to separate causes and consequences in the interactive effects we have described.

The challenge faced by future studies will be to establish a conceptual framework where key mechanisms and responses may be unraveled. An integrated approach is needed to assess the nonnutritional consequences of breastfeeding. The design of such studies should consider assessing not only the effect of psychosocial factors affecting development but also the subtle interactions between mother, infant and the environment. Maternal psychological makeup, family interactions and the environmental stimulation the child receives are key factors that should not be neglected. Case-control studies represent the best next step in this fascinating field of biological research.

The findings that we have summarized may have special relevance for underprivileged communities in industrialized countries and populations in developing countries, since the evidence presented here provides strong reasons to favor human milk feeding. When evaluating early feeding modes we should not only look at growth and biochemical indices but also at functional outcomes, including mental development and other measures of central nervous system function. Breastfeeding is best unless proven otherwise; the burden of proof should be placed on those proposing that bottle feeding can equal human milk from the breast.

References

1 Bowlby J: La perdida afectiva. Buenos Aires, Paidós, 1983.
2 Klaus M, Kennel J: Care of the parents; in Klaus MH, Fanaroff AA: Care of High-Risk Neonates, ed 2. Philadelphia, Saunders, 1979.
3 Klaus MH, Kennel J: Labor, birth and bonding; in Klaus MH, Kennel J (eds): Parent-Infant Bonding, ed 2. St. Louis, Mosby, 1982, pp 21–98.
4 Gunther M: The new mother's view of herself. Ciba Found Symp 1976;45:00.
5 Kennel K, Dowling S, Kennel J: Feeding and behavior: Three recent observations; in Bond JT, et al. (ed): Infant and Child Feeding. New York, Academic Press, 1981.
6 Cunningham N, Anisfeld E, et al: Infant carrying, breast feeding and mother-infant relations. Lancet 1987;i:379.
7 Jellife DB, Jellife FP: Human Milk in the Modern World. Oxford, Oxford University Press, 1978, pp 142–160.
8 Porter M: Attractiveness of lactating females' breast odors to neonates. Child Dev 1989;60:803–810.
9 Tyson J, Burchfield J, Sentance F, et al: Adaptation of feeding to a low fat yield in breast milk. Pediatrics 1992;89:215–220.
10 Orbach S, Eichenbaum L: What Do Women Want? New York, Berkeley, 1978.
11 Bayley Scale of Infant Development, ed 2. Cleveland, Psychological Corporation, 1993.
12 Fagan JF, Singer LT: Infant recognition memory as a measure of intelligence; in Lipsett LP (ed): Advances in Infancy Research. Norwood, Ablex, 1983, vol 2, pp 31–78.
13 Fagan JF III: The relationship of novelty preferences during infancy to later intelligence and later recognition memory. Intelligence 1984;8:339–346.
14 Fagan JF III, Singer LT, Montie JE et al: Selective screening device for the early detection of normal or delayed cognitive development in infants at risk for later mental retardation. Pediatrics 1986;78:1021–1026.
15 Fagan JF III, Shepherd PA, Knevel CR: Predictive validity of the Fagan test of infant intelligence. Proc Soc Res Child Dev Conf, Seattle, 1991.
16 Morgan GA, Harmon RJ, Maslin-Cole CA: Mastery motivation: Its definition and measurements. Early Educ Dev 1990;1:318–339.
17 Terman L, Merrill M: Stanford-Binet Intelligence Scale: Manual for the Third Revision, Form L-M. Boston, Houghton Mifflin, 1960.

18   Anderson CR, Vestal LL, Hicks RM: Laboratory in Educational Assessment and Diagnosis. Special Education Practicum Handbook Series. Anderson, Anfap, 1982.

19   Kirk SA, McCarthy YY, Kork WD: Illinois Test of Psycholinguistic Abilities, rev ed. Los Angeles, Western Psychological Services, 1968.

20   Woodcock RW: Bateria Woodcock Psicoeducativa en Español. Teaching Resources Corporation, 1982.

21   Bruininks R: Bruininks-Oseretsky Test of Motor Proficiency. Circle Pines, American Guidance Service, 1978.

22   Beery KE: Revised Administration, Scoring and Teaching Manual for the Development Test of Visual Motor Integration. Toronto, Modern Curriculum Press, 1982.

23   Parent-Infant Observation Guide for Program Planning. The Ounce of Prevention Fund Developmental Program, 1989.

24   Caldwell BM: Home Inventory for Infants: Instruction Manual, rev ed. Little Rock, University of Arkansas, Center for Early Child Development, 1975.

25   Bradley RH, Ramey CT, et al: Home environment and cognitive development in the first 3 years of life: A collaborative study involving six sites and three ethnic groups in North America. Dev Psychol 1989;25:217–235.

26   Devins GM, Orme CM: Center for Epidemiologic Studies Depression Scale; in Keysas DJ (ed): Test Critique II. Kansas City, Test Corporation of America, 1985.

27   De Andraca I, Cobo C, Rivera F, et al: Evaluación de la Inteligencia a través de formas cortas del WAIS para grupos de población de nivel socioeconómico bajo. Rev Saude Publica 1993;27:334–339.

28   Rodgers B: Feeding in infancy and later ability and attainment: A longitudinal study. Dev Med Child Neurol 1978;20:421–426.

29   Rogan JW, Gladen BC: Breast feeding and cognitive development, Early Hum Dev 1993;31:181–193.

30   Ferguson DM, Beautris AL, Silva PA: Breast-feeding and cognitive development in the first seven years of life. Soc Sci Med 1982;16:1705–1708.

31   Morrow-Tlucak M, Haude RH, Ernhart CB: Breastfeeding and cognitive development in the first two years of life. Soc Sci Med 1988;26:635–639.

32   Jacobson SW, Jacobson JL: Breastfeeding and intelligence. Lancet 1992;339:926.

33   Taylor B, Wadsworth J: Breast feeding and child development at five years. J Dev Med Child Neurol 1984;26:73–80.

34   Bauer G, et al: Breast feeding and cognitive development of three-year-old children. Psychol Rep 1991;68:1218.

35   Jacobson SW, Jacobson JL, Frye KF: Incidence and correlates of breast-feeding in socioeconomically disadvantaged women. Pediatrics 1991;88:728–736.

36   Morley TJ, Cole RP, Lucas A: Mother's choice to provide breast milk and developmental outcome. Arch Dis Child 1988;63:1382–1385.

37   Lucas A, Morley R, Col TJ, et al: Breastmilk and subsequent intelligence quotient in children born preterm. Lancet 1992;339:261–264.

38   Bazan NG: The metabolism of omega-3 polyunsaturated fatty acids in the eye: The possible role of docosahexaenoic acid and docosanoids in retinal physiology and occular pathology. Prog Clin Biol Res 1989;312:95–112.

39   Uauy R, Hoffman DR: Essential fatty acid requirements for normal eye and brain development. Semin Perinatol 1991;15:449–455.

40   Simopoulos AP: Omega-3 fatty acids in health and disease and in growth and development. Am J Clin Nutr 1991;54:438–463.

41   Innis SM: Essential fatty acids in growth and development. Prog Lipid Res 1991;30:39–103.

42   Birch E, Birch D, Hoffman D, et al: Breast-feeding and optimal visual development. J Pediatr Ophthalmol Strabismus 1993;30:33–38.

43   Burr GO, Burr MM: A new deficiency disease produced by rigid exclusion of fat from the diet. J Biol Chem 1929;82:345–367.

44   Hansen AE, Wiese HF, Boelsche AN et al: Role of linoleic acid in infant nutrition: Clinical and chemical study of 428 infants fed on milk mixtures varying in kind and amount of fat. Pediatrics 1963;31:171–192.

Breastfeeding for Optimal Mental Development   25

45   Uauy R, Treen M, Hoffman D: Essential fatty acid metabolism and requirements during development. Semin Perinatol 1989;13:118–130.
46   Caldwell MD, Johnsson HT, Othersen HB: Essential fatty acid deficiency in an infant receiving prolonged parenteral alimentation. J Pediatr 1972;81:894–898.
47   White HB, Turner MD, Turner AC, et al: Blood lipid alterations in infants receiving intravenous fat-free alimentation. J Pediatr 1973;83:305–313.
48   Friedman Z: Essential fatty acids revisited. Am J Dis Child 1980;134:397–408.
49   Holman RT, Johnson SB, Hatch TF: A case of human linolenic acid deficiency involving neurological abnormalities. Am J Clin Nutr 1982;35:617–623.
50   Fleisler SJ, Anderson RE: Chemistry and metabolism of lipids in the vertebrate retina. Prog Lipid Res 1983;22:79–131.
51   Neuringer M, Connor WE, Lin DS, et al: Dietary ω-3 fatty acids: Effect of retinal lipid composition and function in primates; in Anderson RE, Hollyfield JG, LaVail MM (eds): Retinal Degenerations. New York, CRC, 1991, pp 1–13.
52   Wheeler TG, Benolken RM, Anderson RE: Visual membranes: Specificity of fatty acid precursors for the electrical response to illumination. Science 1975;188:1312–1314.
53   Watanabe I, Kato M, Aonuma H: Effect of dietary alpha-linolenate/linoleate balance on the lipid composition and electroretinographic responses in rats. Adv Biosci 1987;62:563–568.
54   Neuringer M, Connor WE, Van Petten C. et al: Dietary omega-3 fatty acid deficiency and visual loss in infant rhesus monkeys. J Clin Invest 1984;73:272–276.
55   Koletzko B, Bremer HJ: Fat content and fatty acid composition of infant formulas. Acta Paediatr Scand 1989;78:513–521.
56   Hansen J: Commercial Formulas for Preterm Infants 1992; in Tsang RC, Lucas A, Uauy R, Zlotkin S (eds): Nutritional Needs of Preterm Infants: Scientific Basis and Practical Guidelines. Baltimore, Williams & Wilkins, 1993, pp 297–301.
57   Farquharson J, Cockburn F, Ainslie PW: Infant cerebral cortex phospholipid fatty-acid composition and diet. Lancet 1992;340:810–813.
58   Uauy RD, Birch DG, Birch EE, et al: Effect of dietary omega-3 fatty acids on retinal function of very-low-birth-weight neonates. Pediatr Res 1990;28:485–492.
59   Hoffman D, Uauy R: Essentiality of dietary omega-3 fatty acids for premature infants: Plasma and red blood cell fatty acid composition. Lipids 1992;27:886–895.
60   Birch DG, Birch EE, Hoffman DR, et al: Retinal development in very-low-birth-weight infants fed diets differing in omega-3 fatty acids. Invest Ophthalmol Vis Sci 1992;33:2365–2376.
61   Birch EE, Birch DG, Hoffman DR, et al: Dietary essential fatty acid supply and visual acuity development. Invest Ophthalmol Vis Sci 1992;33:3242–3253.
62   Koletzko B, Schmidt E, Bremer HJ, et al: Effects of dietary long-chain polyunsaturated fatty acids on the essential fatty acid status of premature infants. Eur J Pediatr 1989;148:669–675.
63   Innis SM, Foote KD, MacKinnon MJ: Plasma and red blood cell fatty acids of low-birth-weight infants fed their mother's expressed breast milk or preterm infant formula. Am J Clin Nutr 1990;51:994–1000.
64   Carlson SE, Cooke RS, Rhodes PG: Effect of vegetable and marine oils in preterm infant formulas on blood arachidonic and docosahexaeonic acids. J Pediatr 1991;120:S159–S167.
65   Dratz E, Deese A: The role of docosahexaenoic acid (22:6 n–3) in biological membranes; in Simopoulos AP, Kifer RR, Martin RE (eds): Health Effects of Polyunsaturated Fatty Acids in Seafoods. New York, Academic Press, 1986, pp 319–351.
66   Salem N Jr, Kim HY, Yergey JA: Docosahexaenoic acid: Membrane function and metabolism: in Simopoulos AP, Kifer RR, Martin RE (eds): Health Effects of Polyunsaturated Fatty Acids in Seafoods. New York, Academic Press, 1986, pp 263–317.
67   Treen M, Uauy RD, Jameson DM, et al: Effect of docosahexaenoic acid on membrane fluidity and function in intact cultured Y-79 retinoblastoma cells. Arch Biochem Biophys 1992;294:564–570.
68   Bourre JM, Francois M, Youyou A: The effects of dietary α-linolenic acid on the composition of nerve membranes, enzymatic activity, amplitude of electrophysiological parameters, resistance to poisons and performance of learning tasks in rats. J Nutr 1989;119:1880–1892.
69   Carlson SE, Carver JD, House SG: High fat diets varying in ratios of polyunsaturated to saturated fatty acid and linoleic to linolenic acid: A comparison of rat neural and red cell membrane phospholipids. J Nutr 1986;116:718–726.

70   Connor WE, Lin DS, Neuringer M: Is docosahexaenoic acid (DHA, $22:6$ n-3) content of erythrocytes a marker for the DHA content of brain phospholipids? FASEB J 1993;7:152A.
71   Birch EE, Birch DG, Petrig B, et al: Retinal and cortical function of infants at 36 and 57 weeks postconception. Clin Vis Sci 1990;5:363–373.
72   Uauy R, Birch E, Birch D, et al: Visual and brain function measurements in studies on n-3 fatty acid requirements of infants. J Pediatr 1992;120:S168–S180.
73   Hoffman DR, Birch EE, Birch DG, et al: Effects of ω-3 long-chain polyunsaturated fatty acid supplementation on retinal and cortical development in premature infants. Am J Clin Nutr 1993;57:807S–812S.
74   European Society of Paediatric Gastroenterology and Nutrition Committee on Nutrition: Comment on the content and composition of lipids in infant formulas. Acta Paediatr 1991;80:887–896.

Ricardo Uauy, MD, PhD, INTA University of Chile, Casilla 138-11, Santiago (Chile)

Simopoulos AP, Dutra de Oliveira JE, Desai ID (eds): Behavioral and Metabolic
Aspects of Breastfeeding. World Rev Nutr Diet. Basel, Karger, 1995, vol 78, pp 28–54

....

Breastfeeding Kinetics

A Problem-Solving Approach to Breastfeeding Difficulties

Verity Livingstone

Vancouver Breastfeeding Centre, Department of Family Practice,
University of British Columbia, Vancouver, B.C., Canada

Contents

Introduction

Breastfeeding has been recognized as the optimum way to nourish and nurture young children. It has proved to be the most cost-effective, health-promoting, and disease-preventing activity mothers can do because it is pivotal to infant growth, development, immunization and child spacing. During the first year of life, infants undergo a rapid rate of growth and development that is unsurpassed at any other period in their lives. In view of this, the World Health Organization recommends that infants should be exclusively breastfed for the first 4–6 months and weaning should occur in the second year of life and in disadvantaged societies even into the third year of life [1].

Over the last two decades, there has been an increase in the number of young mothers wishing to breastfeed, but despite the widespread promotion of breastfeeding and the encouraging increase in its initiation rate, the number of mothers who exclusively breastfeed is <30% in most industrialized countries [2]. As the prevalence of breastfeeding increases, so does the incidence of breastfeeding difficulties. The rate of abandoned breastfeeding has remained high and the average duration of breastfeeding is still only a few weeks. Most infants are started on breast milk substitutes in hospital, shortly after birth [3].

Mothers express a desire to breastfeed but their expectations are so unrealistic or their experience so unsatisfactory that they abandon breastfeeding prematurely, failing to meet their own breastfeeding goals. New mothers often lack the necessary knowledge and skills to overcome minor problems in the first few days or weeks and the workplace is not conducive to maintaining lactation during separation [4].

A number of factors have been found to predict breastfeeding failure, including low infant birth weight, delay in initiating the first breastfeeding experience, parity, maternal age, marital status, previous breastfeeding experience, lack of family and social support, ethnic origin, smoking behavior and predetermined plans for the duration of breastfeeding. Young, single women with low socioeconomic status in industrial settings are the least likely to breastfeed successfully [5–8].

When looking at actual reasons for terminating breastfeeding, a highly consistent pattern emerges. A large proportion of mothers stop breastfeeding early, primarily because of problems associated with initiating and maintaining lactation or technical difficulties with breastfeeding. Many mothers assume they do not have enough milk to satisfy their babies. This belief is often supported by health professionals, family and friends who readily recommend breast milk substitutes as an appropriate solution to their problems, implying that they are an equal alternative; unfortunately, breastfeeding is abandoned without concern. Later weaning is predominantly the result of other determinants such as the mother's return to work, her convenience and a belief that breast milk is no longer necessary to the infant's diet [9].

Society and the health care system are in part to blame for this short duration of breastfeeding. Most breastfeeding failures are directly related to obstacles placed in the way of the mother and baby. Childbirth without intervention is less common, mothers and babies are often separated soon after birth, 24-hour rooming or bedding in is not universally available, prelacteal and complimentary foods are offered without compunction and the public display of breastfeeding is abhorrent to many people [10]. Children are not reared in an environment where breastfeeding is the norm and cultural beliefs may not condone breastfeeding toddlers. Many people naively believe in the virtues and equality of formulas and minimize or disregard the unique active biological properties of human milk.

Traditionally, societies supported new mothers and passed on the skills of parenting including breastfeeding, from generation to generation, through the extended family members. 'Doulas' were female assistants who offered practical advice and support for breastfeeding mothers. In present-day, modern or urban societies, the doula is missing. The advice givers do not have traditional wisdom and lack current scientific knowledge, and what advice they give is often inaccurate, impedes lactogenesis (the initiation of lactation) and prevents successful galactopoiesis (ongoing milk production). Bottles of formula are prescribed, as the answer to breastfeeding difficulties, usually without the mother's informed consent. This undermines her confidence and aggravates the breastfeeding difficulties, resulting in early breastfeeding failure.

To increase the duration of breastfeeding, health professionals must now assume the role of the doula. Primary health care workers are in a key position to offer advice and assistance to breastfeeding families. They must be able to promote breastfeeding prenatally, offer accurate information and screen for lactation difficulties prenatally, provide anticipatory guidance in hospital and accurately diagnose and manage breastfeeding problems postpartum.

Presenting Complaints

Lactation is a physiological process under neuroendocrine control. Most lactation problems are preventable; their etiology is often iatrogenic due to impeded establishment of lactation or inadequate stimulation and drainage of the breast. *Breastfeeding* is a technical process by which milk is transferred from the maternal breast to the infant. Most breastfeeding problems are due to lack of knowledge and technical skills amongst mothers and health professionals.

Lactation and breastfeeding difficulties present in many guises including infant failure to thrive or colic, early introduction of supplements, or maternal breast discomfort such as sore cracked nipples, engorgement or mastitis. These should be considered as symptoms and signs, but not diagnostic. Different clinical syndromes or complexes of symptoms and signs reflect the normal variations in maternal lactation ability and infant breastfeeding ability. For example, the presenting complaints of a mother who has an abundant milk supply and an infant who has a weak suckle will differ from those of a mother whose milk supply is poorly established and who has an infant with a weak suckle.

As in any other area of medicine, problem solving starts with a detailed history and physical examination of both mother and infant, including breastfeeding history and observation. Once the etiology and pathophysiology of the problem have been elucidated, it is relatively simple to offer an effective management plan. Appropriate management of breastfeeding complaints depends on a clear understanding of breastfeeding kinetics.

Breastfeeding Kinetics

The concept of breastfeeding kinetics conveys the idea that breastfeeding is a dynamic interaction between two people, over time. Each phase and factor plays an important role. Minor problems alone may not lead to difficulties, but in combination they may result in breastfeeding failure [11].

Lactation and breastfeeding are processes influenced by multiple factors and forces. Infant growth depends on maternal biopsychosocial health, adequate mammogenesis, unimpeded lactogenesis, successful galactopoiesis and milk production, effective milk transfer and an appropriate quality and quantity of milk intake (fig. 1).

These phases are influenced by multiple facilitating and/or impeding factors: childbirth, stimulation and drainage of the breast, milk ejection, maternal and infant breastfeeding technique, frequency and duration of suckling, and pattern of breast use. These in turn are influenced by other factors such as maternal knowledge, attitude, motivation, mood and health, support from family, friends, and health care professionals, and also by infant behavior and health.

Successful management of breastfeeding difficulties depends on an appreciation of this mother-infant relationship as well as knowledge about the anatomy of the breast, the physiology of lactation and infant suckling mechanics.

Established lactation or ongoing milk production occurs following adequate mammogenesis, unimpeded lactogenesis and continuous galactopoiesis. It is dependent on adequate breast stimulation and regular complete breast drainage. A problem in any one of these areas may lead to decreased milk production. 'The Art of Successful Breastfeeding – A Guide for Health Professionals' is available on

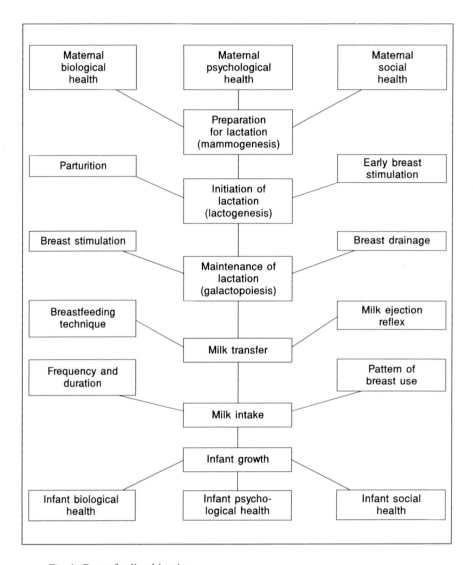

Fig. 1. Breastfeeding kinetics.

video from: Vancouver Breastfeeding Centre, 690 West 11th Avenue, Vancouver, BC V5Z 1M1, Canada, Fax: (604) 875 5017.

Mammogenesis

Intrauterine mammary gland development in the human female starts in the 6th week of gestation with an ectodermal ridge at the site of the glands. Vascularization and formation of specific apocrine glands (Montgomery glands) occurs by 15 weeks and by 8 months canalization is complete and differentation takes place. Following birth, the placental sex hormones may stimulate neonatal breast secretion of milk which subsides spontaneously and the small mammary disc of childhood remains until puberty. With the onset of hypothalamic maturation in the female at 10–12 years, the ovarian graafian follicles mature and initiate the secretion of estrogen which stimulates development of the mammary ducts. The volume and elasticity of the connective tissue surrounding the ducts increase and vascularization and fat deposition are enhanced. The breasts enlarge and mature further under the combined effects of estrogen and progesterone. Differentiation of the breast tissue takes place primarily during adolescence but continues during adult life.

The mature breast contains 15–25 segments or lobes of glandular tissue, surrounded by connective tissue. Not every lobe is functional in each lactation or for the duration of any one lactation; some lobes regress sooner than others. With each subsequent lactation, functional glandular tissue generally increases; the maximum potential for milk production far exceeds the needs of a singleton infant. Milk is secreted in tiny sac-like alveoli, in which there are 10–100 clusters in each segment, enveloped in collagen sheaths. These sheaths form small ducts to convey the milk to the main lactiferous ducts. Beneath this sheath is a lining of contractile myoepithelial cells which contract under the influence of oxytocin from the posterior pituitary and eject milk from the alveoli into the ducts.

The main lactiferous ducts become distensible and enable milk to collect between feeds particularly in the area beneath the areolar tissue. Several milk ducts merge before reaching the nipple. The number of nipple pores does not correspond to the number of breast lobes. Sometimes many ducts open into one large nipple pore. The nipple is located in the center of the pigmented areola which may act as a visual locater for the infant. The size and shape of both nipples and areolar tissue vary considerably; adult nipples may protrude 0.1–1 cm, remain flat or invert. Areolar tissue may be heavily pigmented and occupy half the breast, or may have minimum pigmentation. The areola contains apocrine or Montgomery glands which serve as lubricating and sense organs. The shape of the breast varies from woman to woman. Racial variations may be associated with discoidal, hemispherical, pear-shaped or conical forms.

In early pregnancy, the nipple-areola complex becomes increasingly sensitive, the areola may increase in diameter and the Montgomery tubercles become more prominent. Their sebaceous secretion possesses antibacterial properties. The nipple tissue becomes more protuberant and graspable. The placental lactogens as well as the luteal sequential sex steroids stimulate considerable glandular development. Prolactin levels increase and contribute to the growth and development of the breast. Within the first few weeks of pregnancy, the breasts visibly enlarge. However, as glandular development occurs, fat stores within the breast may be mobilized and the net effect may result in minimal breast changes.

Objective evidence of successful mammogenesis is breast enlargement accompanied by increased breast sensitivity and secretion of early colostrum at the end of the pregnancy. Lack of breast or nipple changes by the end of pregnancy may indicate a problem with mammogenesis and subsequent milk product [12].

Factors that Help Mammogenesis

Preparation for breastfeeding begins prenatally with a discussion about infant feeding. Many women have thought about it, even before pregnancy, but they may be indecisive. It is important to explore women's beliefs, experiences and expectations: their basic information is often inadequate and fallacious or past experiences were unsuccessful. Health professionals must provide information about early infant nutrition, growth and development that is accurate and understandable. The information concerning infant feeding in most textbooks and patient information booklets is often inaccurate and subtly undermines successful breastfeeding. This is particularly true of information provided by infant formula companies. Information should be offered in many different ways: one-to-one discussion, childbirth classes, simple written materials, and visual aids such as pictures, posters and video films.

Expectant mothers should be encouraged to attend prenatal breastfeeding classes and prepare to breastfeed by learning the principles of good breastfeeding technique. Their visual image of a suckling infant at the breast is often derived from aesthetically pleasing pictures in books, pamphlets and from the few relatives and friends who bravely breastfeed in public. The idealistic image of 'a little bundle, held in the crook of an arm, smiling up at mum, while turning towards a peak-a-boo breast' does not convey reality.

Prenatal classes provide an excellent opportunity to offer anticipatory guidance and help prevent subsequent difficulties. Women can learn about lactation and the art of breastfeeding. They can practice sitting comfortably and putting the baby to the breast using a doll supported on a cushion. It is better to learn the principles of successful breastfeeding before childbirth because mothers have considerable mental difficulty grasping new concepts in the first few days postpartum [13–15].

Table 1. Maternal factors associated with short duration of breastfeeding

---

Lack of knowledge, motivation, and/or support
Previous breastfeeding difficulties
Unusual nipple or breast anatomy, lack of elasticity, breast surgery
Maternal use of prescription and nonprescription drugs
Significant maternal illness including endocrine imbalances
Obstetrical complications
Gestational diabetes
Adoption
Depression
Substance abuse
HIV +
Single mother
Teenage mother
Low socioeconomic status
Illiteracy

Any mother with one or more of the above factors is 'at risk' for premature termination of breastfeeding.

There are many culturally dictated methods of preparing nipples, ranging from rubbing with harsh materials to soaking in potent solutions designed to toughen the skin and stretching exercises. The belief is that they will prevent sore nipples. There is little evidence to support the concept of nipple preparation; normal physiological changes of pregnancy make nipple/areolar tissue more graspable, and it is unlikely that stretching exercises alter this. In all likelihood, all they do is focus attention on the breasts and prepare women emotionally towards their future practical function. Nipples and areolar tissue should be soft and pliable. Soreness is a mechanical problem that is avoided by correct attachment to the breast. Some nipples remain flat or inverted which can lead to technical difficulties with attachment. Hard breast shells which are worn inside a brassiere for a few weeks in the last trimester, to encourage the nipple to protrude, may improve their distendability but this is unproven. The physical act of suckling probably alters the nipple shape more effectively. Contrary to the recommendations of most textbooks, soaps can be used regularly during normal bathing – they are not harmful to the skin.

Prenatal Lactation Assessment

During prenatal visits, a lactation assessment should be performed to identify risk factors that might impede successful lactation and breastfeeding (see tables 1, 2). A breast examination should be done in the last trimester, at the end of mammogenesis. Lack of breast enlargement, unusual looking breasts, areolae or

---

Table 2. Infant factors associated with short duration of breastfeeding

Prolonged maternal-infant separation
Admission to hospital or intensive-care nursery
Preterm
Low birth weight
Multiple birth
Congenital anomalies that interfere with suckling, digestion or respiration
Low Apgar score
Significant infant illness
Jaundice

Any infant with one or more of the above factors is 'at risk' for premature termination of breastfeeding.

nipples, and previous breastfeeding difficulties should be considered 'high risk' indicators for lactation insufficiency. Other compounding factors include maternal biopsychosocial factors such as lack of maternal motivation associated with lack of support or knowledge [16].

Factors Interfering with Mammogenesis

The incidence of mammogenesis failure is unknown. Tubular or hypoplastic breasts may not function normally and an obvious discrepancy in breast size may indicate unusual anatomy or physiology. Maternal health factors which interfere with lactation ability include breast surgery. Reduction mammoplasty may disrupt the ducts and nerves and augmentation occassionally interferes with lactation by interrupting the afferent nerve pathway. It must be remembered that abnormal breasts requiring surgery may never function normally because of the underlying problem. Other factors that impede successful mammogenesis may include genetic predetermining phenomena, inadequate breast tissue receptors or an inadequate hormonal milieu. This area has not been studied in detail and many questions remain unanswered [17, 18].

Lactogenesis

Lactogenesis, or the initiation of lactation, occurs close to parturition. It is under endocrine control by the pituitary gland via prolactin and other lactogenic hormones. The decline of placental hormones following the delivery of the placenta, associated with early and frequent suckling, are the major triggers to establishing a good milk supply. Placental remnants may interfere with lactogenesis. If the placental tissue is not fully removed, lactation fails to become fully established and an adequate supply of milk may never occur [19–21].

Soon after delivery, infants exhibit a natural locating reflex and can find the nipple themselves, if they are allowed. Once the nipple is located, they root and latch onto the breast and suckle instinctively. Studies have shown that the suckling instinct can be impaired if foreign objects are inserted into infants' mouths soon after birth [22].

This early suckling is critical for three reasons. First, it allows an imprinting to occur as the infant learns to suckle effectively while the breast is still soft and graspable. Second, the infant ingests a small amount of colostrum which has a high content of secretory IgA, which acts as the first immunization to the immuno-immature infant. Third, early suckling stimulates the release of prolactin and other pituitary hormones which are required to initiate the production of milk [23, 24].

Following delivery, infants should start suckling on the breast as soon as possible and then have unlimited access to both breasts. This causes surges of prolactin from the anterior pituitary resulting in the production of colostrum and then mature milk [25].

Clinical signs of successful lactogenesis are fullness of the breasts and production of colostrum and then milk. Colostrum is present from birth whereas mature milk production starts at approximately 36 h postpartum.

There is a window period for the initiation of lactation. Studies show that the duration of lactation correlates inversely with the time of first breast stimulation; lactogenesis, and hence successful galactopoiesis, is impaired by its delay [26]. This delay commonly occurs when mothers and infants are separated because of existing or anticipated health problems [27].

Factors that Help Lactogenesis
The goal in hospital should be to establish a milk supply and every effort must be made to do this. The joint WHO/UNICEF statement 'Protecting, Supporting and Promoting Breastfeeding' outlines ten simple steps designed to protect this delicate physiological process [28].

Most mothers and infants should not be separated unnecessarily. Bedding in is superior to rooming in; it allows the baby unlimited access to the breast. Mothers who have had cesarean-section birth can enjoy having their babies tucked up beside them. The best observers of newborn babies are their parents.

Parenting starts at birth, hence the hospital staff should encourage mothers to assume this role as soon as possible. Infants instinctively know how to locate the breast and suckle but mothers must be taught. Hospital stays are very short and new mothers need help to breastfeed at every opportunity and must be encouraged to practice under supervision. Nighttime feedings are a reality and offer nurses an excellent opportunity to provide anticipatory guidance and practice preventative medicine. If a mother has to wait until she is alone, at home, she may feel very inadequate, but if she can gain breastfeeding skills during her hospital stay she will be more likely to succeed afterwards [29].

Factors that Impair Lactogenesis

Initiating a milk supply and successful breastfeeding versus resorting to artificial feeding in hospital, result from a delicate balance between receiving good and bad help. Infants instinctively know how to suckle and will competently latch and suckle when they are positioned appropriately or allowed to root for the breast. They can learn to breastfeed over time, if necessary, but the maternal physiological ability to lactate rapidly declines if both breasts are not stimulated and drained every 2–3 h. Lack of understanding of the physiological process of lactogenesis is the root of the problem. Many labor and delivery nurses and doctors do not have the technical skills and knowledge to place a newborn infant correctly at the breast. Some believe in a hands-off approach, allowing a mother 'to figure it out' herself. Unfortunately, breastfeeding is a learned skill, which must be taught.

An inadequate milk supply is the commonest reason for stopping breastfeeding in the early weeks. The cause is often iatrogenic due to mismanagement during the critical early phase. Many hospital policies and practices interfere with the physiological processes of establishing milk production. Hospital confinements influence mothers' behavior with respect to early infant feeding in several respects. The institutional context dictates when and how often the mother is able to see, touch and suckle her baby. Many routines provide mothers with messages about alternative methods of feeding and may promote confusion. Hospital practices often expose mothers to medications and procedures which may make it difficult for her to establish lactation or the baby to breastfeed. There may be staff attitudes which are ambiguous, contradictory and occasionally harmful. It is paradoxical that women who rely most on the medical establishment for infant feeding are the least likely to breastfeed successfully [30–32].

Following delivery there is considerable vascular and lymphatic congestion in the breast tissue leading to a rise in intraductal pressure. If unrelieved, the engorgement impedes the flow of milk in the ducts and reduces circulation which rapidly causes pressure atrophy at the alveoli precluding establishment of a good milk supply [33, 34]. Inadequate galactopoiesis or ongoing milk production is a direct result of impeded lactogenesis. Early and frequent breast drainage is the most effective way of preventing this.

The fluid requirements of healthy newborn infants are minimal for the first few days. Infants ingest 7–20 ml colostrum per feed initially, and do not require extra fluids. Prelacteal and complimentary feeds may upset the process of lactogenesis by removing the infant's hunger drive and decreasing the frequency of breast stimulation and drainage [35, 37].

Breast fullness is common in the first few days; it may prevent infants from latching effectively. This leads to sore nipples, caused by tongue trauma, inadequate drainage and insufficient milk intake. If an infant's breast milk intake is insufficient, he or she remains hungry and may receive formula supplement. The net result is impeded lactogenesis and maternal unhappiness.

Night sedation may offer a mother temporary rest, but lack of breastfeeding at night can also impede lactogenesis due to irregular breast stimulation and drainage.

If frequent efficient breastfeeding is not possible, e.g. if a mother is separated from her preterm infant, she should be shown how to express her milk, either by hand or by regularly using a breast pump, to ensure complete breast drainage and prevent milk stasis. Contrary to popular belief, this does not lead to an excessive milk supply but prevents early and irreversible involution. Studies have shown that a minimum of 100 min pumping per 24 hours are needed to maintain adequate prolactin levels. Manual expression of milk before latching helps to improve the attachment [38]. Pituitary insults following severe hypotension associated with obstetrical hemorrhage (Sheehan syndrome) may result in impaired lactogenesis and hence low milk production. Delay of several days in 'milk coming in' may predict subsequent inadequate milk production. Serum prolactin levels should increase several-fold following suckling; lack of a prolactin response may be significant [39].

Galactopoiesis

Galactopoiesis is the ongoing production of milk. It follows successful mammogenesis and unimpeded lactogenesis, is controlled by regular and complete breast drainage and is primarily an apocrine action. Breast stimulation triggers prolactin and oxytocin release. Prolactin surges stimulate the breast alveolae to actively secrete milk, and oxytocin causes the myoepithelial cells surrounding the glands and the ductules to contract and eject milk down the ducts to the nipples. These contractions effectively squeeze the fat globules across the cell membrane into the ducts. As the feed progresses the quality and quantity of milk produced changes. The fore milk at the beginning of the feed is mainly composed of milk that has collected between feeds. Fore milk has a low fat and higher whey content. The fat content increases during a feed but the volume of milk ingested decreases [40]. The rate of milk synthesis varies between mothers, and partially depends on the frequency of feeds and the removal of local suppressor peptides.

Factors that Help Galactopoiesis

Despite successful initiation of lactation in the hospital, there is still a high incidence of early breastfeeding difficulties. Postpartum support and breastfeeding management advice in the community are vital.

Sometimes parents may not be able to interpret the infant's cues correctly and either under- or overfeed the child. Many well-meaning grandmothers and friends rely on their personal experiences and encourage bottle feeding as a solution to all feeding difficulties. The breast-feeding dyad is at an extremely vulnerable stage at this time; lack of knowledge, lack of confidence and lack of support culminate in breastfeeding abandonment. The doula is critical and must have a high profile during the early postpartum phase. Once mothers have overcome early parenting difficulties, they can successfully breastfeed for as long as they wish.

Factors that Impair Galactopoiesis

Studies suggest that ongoing milk production is inhibited by the buildup of local suppressor peptides; regular suckling removes this inhibition [41, 42]. Incomplete drainage of the breasts because of infrequent suckling or ineffective breastfeeding techniques can result in decreased milk production and early weaning. Engorgement and milk stasis is usually iatrogenic and the result of infrequent breast drainage often due to maternal and infant separation. Infrequent breast stimulation fails to trigger sufficient prolactin surges to maintain lactation, and involution begins.

Infants appear to take control of milk intake during the first months of life and it takes several weeks before breastfeeding is properly established. In the early weeks of life, the infant's daily milk requirements increase quickly. In order for the maternal milk supply to meet this demand, the infant must nurse frequently; the breasts must be emptied regularly [43, 44]. Many mothers state they have been counselled by health professional to introduce their infants to a supplemental bottle at an early age. As a result, the delicate symbiotic relationship of the nursing dyad may be upset and a downward cascade towards premature weaning may begin [45]. The reasons given for introducing this bottle are varied. The most common is that a mother has encountered difficulties with breastfeeding. The belief still exists that bottle feeding is an acceptable alternative to breastfeeding and that there is no justifiable reason to dissuade a mother from her decision to terminate breastfeeding prematurely. Instead of diagnosing the cause of the problem and managing it appropriately, some professionals assume that breast-feeding is proving unsuccessful and in a misguided attempt to help, suggest that bottle feeding would meet the nutritional needs of the infant and put an end to the problems.

Other mothers have stated that they have introduced a relief bottle once a day as an alternative to breastfeeding. It is tempting to ask relief for whom? What exactly is the purpose of this bottle? Is it to provide a time for father to free mother from the chore of feeding and to participate in the nutritive as well as the nurturative aspects of child rearing or is it the introduction of a menace?

If a bottle of formula is substituted for a feed of breast milk during this early learning time, it can significantly contribute to premature termination of breast-feeding. The decreased nipple stimulation and the incomplete emptying of the breasts contribute to the failure to increase milk production and can lead to an insufficient-milk syndrome. This can reinforce a new mother's doubt of her own ability to produce enough milk to meet the demands of her rapidly growing infant and she may resort to giving the infant a bottle more often.

Milk Transfer

The rate of transfer of milk from the breast to the infant is dependent on various factors including milk production and the volume of pooled milk, the strength and frequency of the milk ejection reflex and the technical process of breastfeeding [46]. The milk ejection reflex, or 'let down', is stimulated by oxytocin released from the posterior pituitary following direct nipple stimulation and via hypothalamic triggering. It causes smooth muscle contractions and propels milk through the ducts and out of the nipple pores. Breastfeeding must be observed in order to assess suckling mechanics and milk transfer.

Breastfeeding
Breastfeeding is a technical process of transferring milk from the breast. It is an interactive process. Efficient breastfeeding depends on careful positioning and attachment of the infant onto the breast and on an intact suckling ability of the infant [47].

Positioning. The mother should be sitting comfortably with her arms and back supported and her feet raised on a small stool. The baby should be placed on her lap, facing towards her uncovered breast. A pillow may help raise up the baby. Breastfeeding is easier if two hands are used to start with. The breast should be cupped with one hand using all the fingers underneath and the thumb above, lifting it up slightly while directing the nipples towards the baby's mouth. The other hand is used to support the baby's shoulders and back of the head. The baby's arms should be free to embrace the breast and its body held very close to the mother, stomach to stomach.

Attachment. The latching technique involves brushing the nipple against the baby's mouth and waiting until the mouth opens wide. This often requires 'teasing the baby' and encouraging the mouth to open wider than before. When the mother can see the gaping mouth, she should quickly draw the baby forward over the nipple and areolar tissue and then maintain this two-handed hold throughout the feed. Young infants do not have the ability to maintain their position at the breast alone and so the mother must continue to support her breast and the baby throughout the duration of the feed. Older infants are able to maintain and latch themselves more easily and nurse comfortably in an elbow crook.

Suckling. An infant who is correctly latched and has a mouthful of soft breast tissue, will draw the nipple and the areolar tissue to the junction of the hard and soft palate to form a teat and then initiate suckling. Ultrasound studies show how the gums and jaws are positioned over pressure receptors in the lactiferous ducts and trigger milk ejection. The jaw is raised and the gums compress the lactiferous sinuses under the areola, the tongue protrudes over the lower gums, grooves and undulates, in a coordinated manner, to strip out milk from the teat. The jaw lowers and the soft palate elevates to close the nasopharynx, a slight negative pressure is created and the milk is effectively transferred and swallowed [48].

Factors Impeding Milk Transfer

The milk ejection reflex is a primitive one and is not easily blocked. Maternal anxiety may induce a stress reaction with the release of adrenalin which can cause vasoconstriction and impede the action of oxytocin but over time this inhibition seems to be overcome. The strength and frequency of the ejection reflex to hypophysial stimulation of the posterior pituitary and suckling pressure on the lactiferous ducts causing oxytocin release. The more milk that has pooled between feeds, the more is ejected with the initial let down. The character of this reflex varies between women and over time; some mothers have a well-developed let down, others have a slow irregular reflex. Confidence facilitates the ejection reflex while anxiety may impede it [49].

Some infants have a strong vigorous suckle that triggers an ejection reflex quickly and demonstrate long bursts of nutritive suckling with few pauses. Others, including low-birth-weight infants have weak suckles that do not easily evoke a let down. These babies do considerably more nonnutritive suckling and have shorter bursts of swallowing.

Breastfeeding technique problems may be due to poor maternal technique associated with poor positioning and latching, or to infant problems associated with sucking difficulties. Improper positioning and attachment result in decreased breast stimulation, and inadequate drainage. This results in decreased milk production and decreased milk intake [50]. Simple correction of the position and latch is often the only remedy needed to improve the quality of the feed. When a baby is correctly latched, he or she forms a teat out of the mouth full of breast tissue. The more elastic and extensile the breast tissue, the easier it is for the young infant. A fixed, retracted or engorged nipple and areolar tissue makes it harder for this to occur. Manual manipulation, including gentle pulling and stretching out of the areolar tissue and manual expression of milk before feeds, and the use of hard plastic nipple shells can help improve the protuberance and graspability of the tissue [51].

An infant's suck should be evaluated carefully. Visual and digital examination of the mouth is necessary. An examining finger is slowly inserted into the baby's mouth and the tongue should curve gently around it, protruding over the gums. The suckle should be rhythmic, coordinated and equal. Infants with a cleft palate or an uncoordinated, weak, flutter or bunched-up tongue may have an impaired ability to suckle effectively, often because they fail to create the necessary negative pressure to draw in the milk. They require specialized retraining. Ankyloglossia (tongue-tie) is an important cause of suckling difficulties. The tethered tongue is unable to protrude over the gum and cannot move upward; the teat is not stripped correctly and less milk is transferred. The nipple often becomes traumatized and sore. The baby may then not thrive, and milk production decreases because of inadequate drainage. A simple surgical release of the frenulum is required and should be done as soon as possible when clinically indicated; after a few weeks it is often difficult to alter the way these infants suckle [52].

It has been clearly documented that the infant's sucking action at the breast is very different from the sucking action on a rubber nipple. The position of the infant at the breast and the precise sucking action on the nipple alters its shape during feeds so that the human nipple conforms to the infant's mouth; by contrast, the infant's mouth must conform to a rubber nipple. The bottle requires a different technique: the tongue is frequently positioned at the tip of the nipple in order to slow the flow of milk, and the jaws do not compress the nipple. An infant who has learned to suck on a rubber nipple may become confused at the breast. Studies have shown that the mean latency time for release of milk is 2.2 min after the infant begins to suck at the breast. There is no latency period for bottlefed infants. Some infants become accustomed to the immediate rewards of the bottle and turn away from the breast [48].

To prevent nipple confusion, rubber nipples and pacifiers should be avoided, particularly if the mother's nipples have poor graspability. This inevitably leads to frustration, sore nipples and ultimately the infant's refusal of the breast in favor of the rubber nipple. There are many orthodontic teats on the market. The manufacturers claim that these teats simulate the mother's nipple, but the basis for these claims is unclear. It seems likely that all rubber nipples present the same potential problems and can cause nipple confusion [53].

A common symptom of breastfeeding difficulties is sore nipples. Soreness in the first few days is almost always due to tongue trauma and indicates poor breastfeeding technique. A correctly latched infant will not cause trauma. Persistent sore nipples with fissures or ulcers may require 24 h rest. Lanolin cream and other remedies have not proved effective and should be avoided. The presence of fissures and ulcers on the tip of the nipple in conjunction with a purulent exudate is highly suspicious of a *Staphylococcus aureus* skin infection and usually requires antibiotic treatment. Moniliasis is another cause of soreness that may be overlooked. Oral thrush in the infant is associated with maternal nipple infection; both mother and baby should be treated with antifungal therapy. The infant's mouth should be painted with nystatin solution after feeds and an effective fungal ointment should be applied to the nipples before and after each feed for 10–14 days.

Milk Intake

The growth of a healthy breastfed infant depends on its daily milk intake. Milk intake depends on ongoing galactopoiesis, effective milk transfer, frequency and duration of breastfeeding and the pattern of breast use. A young infant drinks about 760–840 ml of milk per day, usually feeding 8–10 times. The milk intake per feed is about 80–100 ml. Breasts have a great capacity to yield milk and can produce double this amount. If necessary, a woman can feed from one breast exclusively.

Frequency
Infants are able to recognize hunger and should be fed on demand. Most young infants breastfeed every 2–3 h; this causes frequent surges of prolactin which help to ensure full lactation. Mothers who have a low milk supply should be encouraged to breastfeed frequently to ensure good drainage and stimulation.

Duration
Studies show that the duration of a breastfeed varies between mother-infant pairs. The rate of milk transfer is not uniform. Some breastfeeding pairs have a rapid milk transfer and, hence, a very short feed. This is because of the large amount of milk that has collected in the breasts since the previous feed and the well-established milk ejection reflex. Others have long feeds because milk ejection is poor, the breastfeeding technique is relatively ineffective, or milk production is slow and the pooled milk volume is low and, hence, milk transfer is slow. Previously held beliefs that most of the feed was taken in the first few minutes or that both breasts should be used at each feed failed to recognize the uniqueness of each nursing pair [54].

The total volume of milk intake also varies and seems to depend on total fat and caloric intake. Infants recognize satiation and spontaneously stop suckling when they are full. A high-fat, low-volume feed may have a different rate of transfer than the lower-fat, high-volume feed. The exact mechanism of control has not been fully elucidated [55].

Pattern of Breast Use
The quality and quantity of milk intake depends on the pattern of breast use. The first breast must be fully drained before switching to the second. This will

prevent milk stasis, and results in balanced milk production and optimum infant growth.

Mothers with a good milk yield can readily feed from one breast per feed; this ensures that the infant obtains the higher fat content in the hind milk from the first breast and less of the lactose-rich fore milk from the second breast. Mothers with a lower milk yield should switch to the second side according to the infant's cues. Mothers with a poor milk yield must be encouraged to fully drain both breasts at each feed to provide the infant's necessary milk requirement and to build up their supply.

Factors that Help Milk Intake

The infant should be put to the breast that feels the fullest and held there comfortably in a way that allows effective suckling until the nipple is spontaneously released. If, after burping, the infant still exhibits hunger cues, he or she should be put back to finish the first breast before switching to the second side. If the rate of milk transfer is rapid, the infant may gag, choke and pull away from the breast; frequent burping is recommended in this situation, as is manual expression of milk before attaching the baby.

Factors that Impair Milk Intake

Breastfeeds that take more than 45 min are probably inefficient and require careful evaluation. Some mothers terminate a feed prematurely when the infant appears to be sleeping at the breast. Infants frequently pause while feeding and these episodes may last several minutes; the infant is not asleep and should be allowed to remain on the breast. There is no need to use a finger to break the infant's seal on the breast. Problems arise when mothers terminate a feed, or switch to the other side prematurely. This alters the quality and quantity of the feed. The infant fails to drink the high-fat, hind milk at the end of the first-breast feed and may ingest a larger volume of fore milk from the second breast.

Short, frequent feeds may indicate inadequate milk intake and so the breastfeeding technique should be observed; a 'happy to starve' infant that sleeps for long periods may fail to thrive due to inadequate milk intake [55].

Insufficient Milk Syndrome

The commonest reasons for abandoning breastfeeding in the postpartum period are inadequate milk production and inadequate milk intake leading to failure to thrive and early introduction of supplemental feedings. The etiology is multifactorial but most causes of this insufficient-milk syndrome are reversible if the mother receives accurate breastfeeding management advice early in the

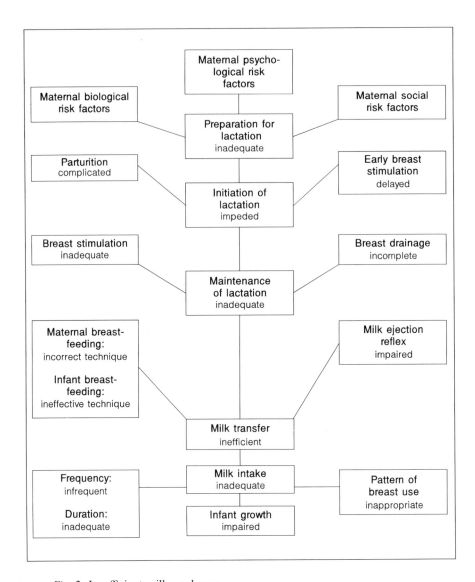

Fig. 2. Insufficient-milk syndrome.

postpartum period. A small percentage are irreversible, often because of failed mammogenesis or impeded lactogenesis. With a move towards early hospital discharge, mothers leave hospital before lactation is established and before they have mastered the technical skills required to breastfeed effectively. Mothers and infants are at risk for developing insufficient-milk syndrome and hence premature

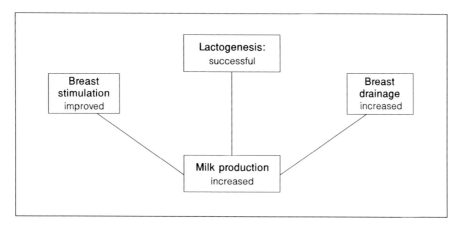

Fig. 3. How to increase milk production.

termination of breastfeeding, unless they receive help from community support services.

Several biopsychosocial risk factors for the insufficient-milk syndrome can be identified prenatally, in hospital and in the early postpartum period by performing a careful lactation assessment. Based on breastfeeding kinetics, the underlying cause is easy to elucidate and management can be directed to either increase milk production, improve milk transfer or increase milk intake (fig. 2).

How to Increase Milk Production

Breast drainage should be improved by ensuring regular, complete emptying of both breasts about eight times in 24 h. After breastfeeding, any residual milk should be removed by manual or mechanical expression; this will ensure complete removal of local inhibitors. Maximal breast stimulation is achieved when the infant is correctly attached and sucks effectively; additional stimulation and hence increased prolactin release can be obtained when the breasts are pumped simultaneously for about 10 min (fig. 3).

Galactagogues including phenothiazine and domperidone are dopamine antagonists and hence remove prolactin inhibition. They may be used as an adjunct therapy in mothers with faltering milk production.

How to Improve Milk Transfer

The art of breastfeeding is a learned skill; mothers need to be taught effective breastfeeding techniques of positioning and latch. Observation is necessary to ensure correct maternal technique. Infants breastfeed instinctively and must be given ample opportunity to suckle at the breast. Objects such as pacifiers and rubber nipples should be avoided.

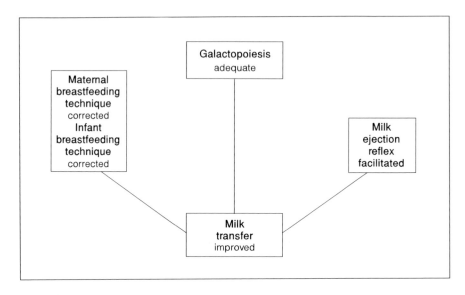

Fig. 4. How to increase milk transfer.

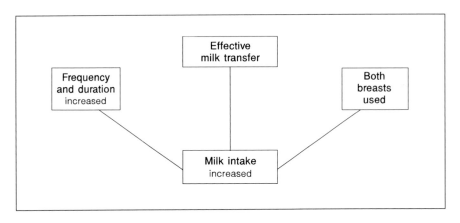

Fig. 5. How to increase milk intake.

The milk ejection reflex is triggered by a correctly attached infant and is strengthened by knowledge and support. Negative stimuli such as pain, lack of confidence or embarrassment should be dispelled (fig. 4).

How to Increase Milk Intake

The amount of milk ingested in 24 h can be increased by ensuring effective milk transfer and including night feeds. The duration of each feed depends on the rate of milk transfer and must be guided by the infant's cues. Both breasts should be offered at each feed (fig. 5).

Complimentary hand feeding may be necessary if the milk intake is inadequate. Expressed breast milk, donor breast milk or breast milk substitutes may be offered after breastfeeding. The additional fluids may be given by a cup, bottle, spoon, eyedropper, supplemental nursing system or via a wet nurse. Each method should be clearly explained to the mother so that she may make an informed decision. The nutritious and nurturing aspects of partial breastfeeding must be emphasized.

Infant Hyperlactation Syndrome

Infants of mothers with abundant milk are often plump and fussy. They tend to choke, gag and pull off the breast, their feeds are short and frequent, and they appear colicky, gassy with explosive watery stools, and have numerous wet diapers. These babies experience a rapid transfer of milk when suckling and may react negatively to the strong milk ejection. After several weeks, the infant may spend less and less time suckling and eventually reject the breast. The initial rapid weight gain is replaced by a slow weight gain. The high lactose content in the fore milk leads to a relative lactose overload and symptoms typical of overfeeding and lactose intolerance [56]. A simple maneuvre that decreases the total quantity of milk and lactose intake and increases the quality of fat intake involves allowing the baby to remain at one breast throughout the feed, encouraging more hind milk intake. This often results in a dramatic improvement for both mother and infant. Complete drainage of one breast increases the energy content of the feed and hence reduces the total number of feedings per day. The infrequent emptying allows the buildup of local suppressor peptides which decreases milk production. The improved drainage of the first breast prevents milk stasis and mastitis (fig. 6).

Maternal Hyperlactation Syndrome

Blocked ducts, milk stasis and abscess formation are all symptoms of incomplete breast drainage. Most mothers experiencing these symptoms have a naturally abundant milk supply and thriving infants, or else they have started to wean and have missed some feeds. The problem often occurs when a mother switches her infant from one breast to the other, before the first side has been adequately drained. If this happens repeatedly, some of the ducts and lobules remain constantly full, leading to a blockage and milk stasis. Mothers complain of 'knife-like' cramps or 'shooting' pains deep in the breast. A firm, lumpy, slightly

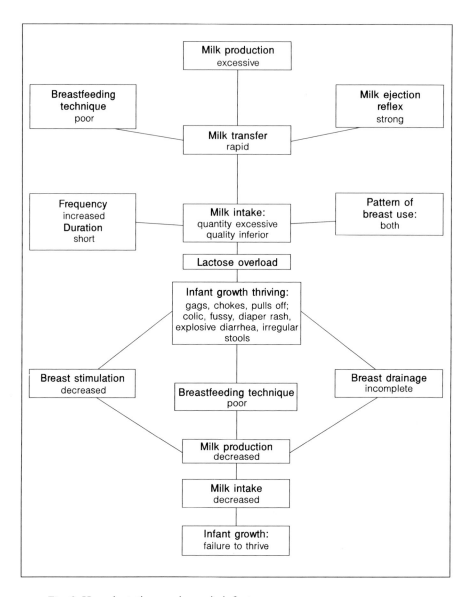

Fig. 6. Hyperlactation syndrome in infants.

tender quadrant in the breast may be felt. Over time, this area becomes inflamed and erythematous, leading to an inflammatory mastitis. The treatment of choice is to drain the breast and unblock the ducts.

This is done by using a correct breastfeeding technique and effective breast milk expression, either manually or with a hand or electric breast pump. If this

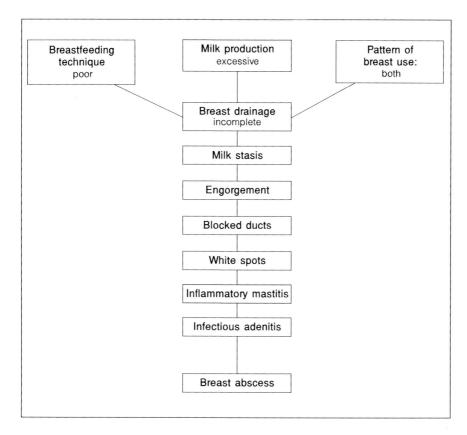

Fig. 7. Hyperlactation syndrome in mothers.

fails to relieve the engorged segment, a technique known as manual stripping may be used. This involves applying firm steady pressure over the breast, starting from the periphery and drawing the fingers and thumb slowly together towards the nipple. The skin must be well lubricated with oil before attempting to do this [57].

A small white dot on the nipple may become visible; this indicates the blocked duct. A sterile needle can be used to lift the epithelial skin off the nipple pore to release the milk. A thick stream of milk will gush out, indicating patency.

An inflammatory mastitis can develop into an infection. If this occurs, pus should be stripped out of the duct using a similar stripping technique. Mothers must be taught how to do this themselves, standing in the shower, using soapy fingers. This is the best method of resolving problems associated with milk stasis. Early antibiotic coverage with erythromycin, cloxacillin or cephalosporin should be considered for 7–10 days if an infection is suspected. *S. aureus* and *Streptococcus* are the common pathogens.

The clinical signs of abscess formation are a high fluctuating fever with chills and general malaise, associated with a firm, well-demarcated, tender, erythematous mass in part of the breast. Ultrasound is a useful diagnostic tool and an aspiration may confirm a collection of pus [58]. Incision and drainage under local or general anesthesia are required, leaving a large drain. The incision should be radial, not circumferential, to minimize duct severance. The dressings should be applied well away from the areola, to allow breastfeeding or expression of milk. Continued frequent breast emptying is crucial to ensure rapid resolution.

The maternal hyperlactation syndrome can be prevented if infants finish the first breast first before switching to the second side and if mothers are instructed to recognize the early signs of milk stasis and ensure the appropriate segment is fully drained in order to prevent recurrence (fig. 7). It is rare for mothers of slow-growing infants to develop mastitis.

Conclusion

When a mother and her infant present with a breastfeeding problem, maternal lactation ability, breastfeeding technique and infant health should be evaluated and specific management offered to correct the underlying cause [59]. Mothers who want to breastfeed need support from health professionals to initiate and establish successful breastfeeding and overcome difficulties.

References

1   Innocenti Declaration: On the Protection, Promotion and Support of Breastfeeding. Document 10017, New York, UNICEF, 1990.
2   Emerg JL, Scholey S, Taylor EM: Decline in breastfeeding. Arch Dis Child 1990;00:369–372.
3   Tanaka P, Leung D, Anderson O, et al.: Infant feeding practices 1984–85 versus 1977–78. Can Med Assoc J 1987;136:940–944.
4   Livingstone VH, Grams GD: Breastfeeding and the working mother. Can Fam Physician 1985; 31:1685–1693.
5   Feinstein JM, Berkel Lamer JE, Gruszka ME, et al.: Factors related to early termination of breastfeeding in an urban population. Pediatrics 1986;78(2):210–215.
6   Sjolin S, Hofvander Y, et al.: Factors related to early termination of breastfeeding: A retrospective study in Sweden. Acta Paediatr Scand 1977;66:505–511.
7   Grossman LK, Fitzsimmons SH, Larsen-Alexander JB, et al.: The infant feeding decision in low and upper income women. Clin Paediatr 1990;29(1):30–37.
8   Goodine LA, Fried PA: Infant feeding practices: Pre- and postnatal factors affecting choice of method and the duration of breastfeeding. Can J Public Health 1984;75:439–443.
9   Bottoff JL: Persistence in breastfeeding: A phenomenological investigation. J Adv Nurs 1990;15: 201–209.
10 Reiff MI, Essock-Vitale SM: Hospital influences on early infant-feeding practices. Pediatrics 1985;76:872–879.
11 Livingstone V: Problem-solving formula for failure to thrive in breastfed infants. Can Fam Physician 1990;36:1541–1545.

12   Infant Feeding: The Physiological Basis. Sci J World Health Organ, bulletin supplement to vol 67, 1989, pp 19–40.
13   Whitley N: Preparation for breastfeeding. J Obstet Gynecol Neonatal Nurs 1978;78:1141.
14   Hewat RJ: Breastfeeding preparing mothers: Prenatal lactation assessment. Br Columbia Med J 1992;34:88–90.
15   La Leche League International. The Womanly Art of Breastfeeding, ed 4. Markham, Penguin, 1987, pp 47–67.
16   Loughlin HH, Clapp-Channing NE, Gehlbach SH, et al.: Early termination of breastfeeding: Identifying those at risk. Pediatrics 1985;75:508–512.
17   Lawrence RA: Breastfeeding: A Guide for the Medical Profession, ed. 3. St. Louis, Mosby, 1989.
18   Neifert M, Demaizo S, Seacat J, et al.: The influence of breast surgery, breast appearance and pregnancy-induced breast changes on lactation sufficiency as measured by infant weight gain. Birth 1990;17:31–38.
19   Aono T, Shioji T, Shoda T, et al.: The initiation of human lactation and prolactin response to suckling. J Clin Endocrinol Metab 1977;44:1101–1106.
20   Willis CE, Livingstone V: Infant insufficient milk syndrome associated with maternal postpartum hemorrhage. J Hum Lactation 1995, in press.
21   Neifert M, McDonough S, Neville M: Failure of lactogenesis associated with placental retention. Am J Obstet Gynecol 1981:140:477–478.
22   Righard L, Alade M: Effect of delivery room routines on success of first breast-feed. Lancet 1990;336:1105–1106.
23   Hanson LA, Adlerberth I, Carlson SB, et al.: Breastfeeding protects against infection and allergy. Breastfeed Rev 1988;13:19–22.
24   Howie PW, et al.: The relationship between suckling induced prolactin response and lactogenesis. J Clin Endocrinol Metab 1980;50:670–673.
25   Kulski JK, Hartman PE: Changes in milk composition during the initiation of lactation. Aust J Exp Biol Med Sci 1981;59:101–114.
26   Salariya EM, Easton PM, Cater JI: Infant feeding: Duration of breast-feeding after early initiation and frequent feeding. Lancet 1978;ii:1141–1143.
27   Elander G, Lundberg T: Short mother-infant separation during week of life influences the duration of breastfeeding. Acta Paediatr Scand 1984;74:237–240.
28   WHO/UNICEF: Protecting, Promoting and Supporting Breastfeeding: The Special Role of Maternity Services. Geneva, World Health Organization 1989.
29   Royal College of Midwives: Successful Breastfeeding, ed 2. Edinburgh, Churchill Livingstone, 1991.
30   Livingstone V: Liberty bottle or liability bottle? A formula for failure. Cam Fam Physician 1988;34:1143–1146.
31   Righard L, Alade MO: Effect of delivery room routines on success of first breastfeed. Lancet 1990;336:1105–1107.
32   Klaus MH, Jerauld R, Kreger NC, et al.: Maternal attachment: Importance of the first postpartum days. N Engl J Med 1972;286:460–463.
33   Moon JL, Humenick SS: Breast engorgement: Contributing variables and variables amenable to nursing intervention. J Obstet Gynecol Neonatal Nurs 1988;00:309–315.
34   Newton M, Newton N: Postpartum engorgement of the breast. Am J Obstet Gynecol 00;61:664–667.
35   Lennon I, Lewis BR: Effect of early complementary feeds on lactation failure. Breastfeed Rev 1987;11:24–26.
36   Shrago L: Glucose water supplementation of the breastfed infant during the first three days of life. J Hum Lactation 1987;3:82–86.
37   Houston MJ: Factors affecting the duration of breastfeeding. I: Measurement of breastmilk intake in the first week of life. Early Hum Dev 1983;8:49–54.
38   Bocar D, Shrago L: Engorgement. Oklahoma City, Lactation Consultant Services, 1990.
39   Sheehan HL, Murdoch R: Postpartum necrosis of the anterior pituitary: pathological and clinical aspects. J Obstet Gynecol Br Empire 1938;45:456–489.
40   Hartmann PE, Kent JC: The subtlety of breastmilk. Breastfeed Rev 1988;13:14–18.
41   Peaker M, Wild CJ: Milk secretion: Autocrine control. News Physiol Sci 1987;2:12406.

42   Prentice A, et al.: Evidence for local feed-back control of human milk secretion. Biochem Soc Trans 1989;17:489–492.
43   Egli GE, et al.: The influence of the number of feedings on milk production. Pediatrics 1961;27: 314–317.
44   Hartmann PE, Prosser CG: Physiological basis of longitudinal changes in human milk yield and composition. Fed Proc 1984;9:2448–2453.
45   Frank DA, Wirtz SJ, Sorenson JR, et al.: Commercial discharge packs and Breastfeeding counselling: Effects on infant-feeding practices in a randomized trial. Pediatrics 1987;86:845–853.
46   Lucas A, Lucas PJ, Baum JD: Patterns of milk flow in breastfed infants. Lancet 1979;ii:57–58.
47   Applebaum RM: The modern management of successful breastfeeding. Pediatr Clin North Am 1977;241:37–47.
48   Woolridge MW: The 'anatomy' of infant sucking. Midwifery 1986;2:164–171.
49   Newton M, Newton N: The let-down reflex on human lactation. J Pediatr 1948;33:698–704.
50   Davies DP: Is inadequate breastfeeding an important cause of failure to thrive? Lancet 1979; ii:541–542.
51   Renfrew M, Fisher C, Arms S: Best Feeding: Getting Breastfeeding Right for You. Berkeley, 1900.
52   Frantz KB, Fleiss PM: Ineffective suckling as a frequent cause of failure to thrive in the totally breastfed baby, in Freix S, Eldelman A (eds): Human Milk: Its Biological and Social Value. Amsterdam, Excerpta Medica, 1980, pp 318–321.
53   Minchin M: Breastfeeding Matters. North Sydney, Allen & Unwin, 1985.
54   Woolridge MW, Baum JD, Drewett RF: Individual patterns of milk intake during breastfeeding. Early Hum Dev 1982;7:265–272.
55   Klaus M: The frequency of sucking: A neglected but essential ingredient of breastfeeding. Obst Gynecol Clin North Am 1987;14:623–633.
56   Woolridge MN, Fisher C: Colic, overfeeding and symptoms of lactose malabsorption in the baby: A possible artifact of feeding management. Lancet 1988;ii:1382–1384.
57   Cantlie HB: Treatment of acute puerperal mastitis and breast abscess. Can Fam Physician 1988;34:2221–2226.
58   Dixon JM: Repeated aspiration of breast abscesses in lactating women. BMJ 1988;297:1517–1518.
59   Ellis DJ, Livingstone V, Hewat R: Assisting the breastfeeding mother: A problem-solving process. J Hum Lactation 1993;9:89–96.

Verity Livingstone, MB, BS, FCFP, IBCLC, Medical Director, Vancouver Breastfeeding Centre, Associate Professor, Department of Family Practice, University of British Columbia, 690 West 11th Avenue, Vancouver, BC V5Z 1M1 (Canada)

Simopoulos AP, Dutra de Oliveira JE, Desai ID (eds): Behavioral and Metabolic
Aspects of Breastfeeding. World Rev Nutr Diet. Basel, Karger, 1995, vol 78, pp 55–73

....
Human Milk and Premature Infants

F.E. Martinez [a], *I.D. Desai* [b]

[a] Department of Pediatrics, University of São Paulo Medical School, São Paulo, Brazil,
and
[b] School of Family and Nutritional Sciences, University of British Columbia,
Vancouver, B.C., Canada

Contents

Introduction

The present knowledge of human physiology and the development of new life
support techniques have permitted significant improvement in the survival rate
and the quality of life for preterm infants, including those of shortest gestational
age and of very low birth weight [1, 2]. As the survival rate improves, the question
of feeding these infants becomes increasingly crucial.

Discrepancies in the literature show that the feeding of preterm and small-
for-gestational-age infants is far from being a fully resolved matter [3–8].
According to the Nutrition Committee of the American Pediatrics Association
[9], ideal feeding for preterm newborn infants in good condition 'supports a rate of

growth approximating that of the third trimester of intrauterine life, without imposing stress on the developing metabolic or excretory systems'. As expected, this definition has not been universally accepted. While, on the one hand, there is not much disagreement that the metabolic system of the infant should not be overloaded, on the other hand, the premise that an attempt should be made to match intrauterine growth has been questioned.

One of the criticisms refers to what should be considered the 'ideal intrauterine growth'. The adequacy of fetal growth is usually evaluated at birth by comparing infant weight and gestational age to the standard curves available for intrauterine growth [10–12]. These standard curves have been based on data obtained in transversal studies which correlated infant weight and gestational age with different gestational ages at birth. Doubts have been raised about the validity of these standards, since, in addition to being transversal, they rest on the basic assumption that prematurely born infants present a growth identical to that of fetuses that remain inside the uterus [13, 14]. In the near future, with the growing sophistication of ultrasound equipment, it may be possible to obtain intrauterine growth curves based on measurements of infants who will be born normally.

Another criticism moved against the definition of the American Academy of Pediatrics is that there is no good teleologic reason to assume that extrauterine growth should be the same as intrauterine growth, given the wide difference between the two environments and the different needs of the organism in the extrauterine medium [14]. The human fetus enjoys a very rapid growth rate in its privileged intrauterine environment. It has no responsibility for maintaining its temperature, working against gravity, digesting and absorbing food, or doing the work of breathing on a regular basis. Its growth rate is restrained only by the size of the uterus and the efficiency of the placental circulation [15]. The extrauterine environment certainly affords no such privileges.

Another questionable point in the definition of the North American Academy concerns growth. In this definition, the standard of dietary adequacy is based on somatic growth. However, in our society, the success of an individual basically depends on intelligence and not on somatic growth. If it is true that adequate growth is associated with better intellectual performance, it is also true that the goal of an often exaggerated growth may require excessive nutritional supplies and lead to, for example, high serum levels of amino acids, which may impair brain growth. Thus, we understand that ideal infant nutrition provides the best possible conditions for the full development of all physical and intellectual potentials granted the child by its genetic heritage.

Although theoretical knowledge about the nutritional requirements of preterm infants has greatly increased over the last few years, this topic is still being debated. There is general agreement, however, about one point – these requirements are greater than those of term infants [16–20]. In fact, the nutritional requirements of these infants are greater than at any time in human life. At no other stage of extrauterine life does body weight triple within 3 months.

Regardless of the real values of these requirements, the fact that they are greater than those of term newborns creates a basic problem. We can easily accept the fact that maternal breastfeeding is the best way to feed healthy term babies [21, 22]. We shall then have to assume that, in view of their greater requirements, this milk may be deficient in several nutrients for preterm infants. We then face the question which has been asked for more than 40 years: is human milk a proper food for such infants [23]?

Importance of Human Milk for Premature Infants

In order to assess the nutritional implications of feeding human milk, we must first distinguish between pooled human milk (which has been used in most of the studies reported) and the child's mother's milk. For pooled human milk, another distinction should be made between the milk of mothers of term infants and 'gestational-age-specific' or preterm 'postpartum-week specific' human breast milk.

Most studies suggest that, at least for the first 2 weeks postpartum, mothers who delivered prematurely secrete milk which is most appropriate for the needs of their infant, i.e., milk with greater concentrations of energy, electrolytes and protein but possibly with the same amounts of vitamins A, C and E and immunoglobulins as the milk of mothers of term infants [24–31]. However, some studies, including one carried out in Brazil, do not confirm these results [32, 33].

These findings have encouraged some centers to feed preterm infants their own mother's milk as the only source of food. The results obtained in these studies have not provided a definitive answer to the question of premature-infant feeding, especially with respect to those of the shortest gestational age and weighing less than 1,500 g at birth.

For some authors, the anthropometric growth of preterm infants fed milk from their own mother is inferior to that of infants receiving special formulas [33], whereas for others it is comparable [3]. At least two points are always raised when the question of feeding a preterm infant its own mother's milk arises: the difficulty in expressing appropriate volumes of milk until the infant can breastfeed directly, and the need for a good support infrastructure for the mother, who will have to continue to satisfy her infant's needs on a daily basis when she is back home. However, there is strong evidence that both maternal milk and special formulas have led to greater growth of preterm infants than banked human milk [34–42].

In addition to lower somatic growth, there are studies showing that preterm newborns exclusively fed human milk manifest symptoms of micronutrient deficiency. Bone mineralization seems to be impaired in infants fed their own mother's milk when compared with infants receiving formula [43]. However, after hospital discharge, bone mineralization problems have also been detected in babies fed formulas primarily created for term infants [44].

Merits of Human Milk versus Special Formulas

The relative merits of special formulas and their own mother's milk for preterm infants have not been fully defined [45–47]. It should be remembered, however, that human milk has advantages that are peculiar to it. In addition to the better bioavailability of its nutrients, human milk has well-known anti-infectious properties [48–51], facilitates digestion [52] and favors intestinal maturation [53]. Other less easily demonstrable advantages have also been pointed out, such as the possible affective importance for the mother who, when expressing her own milk, can feel that she is effectively collaborating in the treatment of her child, in addition to the increased possibility of successful breastfeeding when the infant reaches the appropriate degree of maturation to suckle directly from the breast [54].

Despite the controversies it is our opinion that human milk with a few adaptations can and should be used as the first-choice food for premature infants, including those weighing less than 1,000 g at birth. If there are problems with its physicochemical characteristics, studies should be conducted to adapt human milk to the specific needs of preterm infants rather than abandoning it in favor of formulas also adapted but based on the milk of another species. In this way, many advantages inherent in human milk could be maintained and its possible short-comings could be simultaneously corrected. Compensating for these deficiencies will then be one of the major challenges in preterm-infant nutrition.

Efforts to improve the nutritional quality of human milk began some four decades ago [55, 56]. The lower nutritional density of human milk can be compensated by offering greater milk volumes. Some authors were able to achieve adequate preterm growth by feeding these infants human bank milk and offering volumes of as much as 300 ml/kg/day [57, 58]. This proposal, although simple and effective, has the major limitation of offering milk volumes that may not be tolerated by some infants.

Homogenization of Human Milk

Due to the immaturity of suckling and deglutition presented by preterm infants born before the 34th week of gestation [59], they are fed milk through a gastric tube. When the option to use human milk is made, it will need to be processed before storage, and the procedures will depend on the care taken to collect the milk and on the interval between expression and use [60]. All of this manipulation changes some of the special properties of human milk [61, 62]. One of the demonstrated alterations is a consequence of the instability, in solution, of the fat of expressed milk. When human milk is administered through a tube, losses of as much as 47% in fat content have been reported [63–68].

Fat represents approximately 50% of the caloric content of human milk, and both human and cow's milk contain, in their cream phase, nutrients such as iron, copper, zinc, calcium and magnesium in addition to liposoluble vitamins [69–71]. It may thus be inferred that significant nutritional impairment will occur with fat losses. This fact is enough to explain the disappointing results of many studies on the growth of preterm infants fed human milk. If the objective is to feed human milk to infants through a gastric tube, efforts should be made to avoid, or at least to minimize, the loss of nutrients that may occur during the process.

Human milk lipids are distributed into three populations of fat globules. A population of small globules with a diameter <2 µm stays dispersed for long periods, whereas globules of larger diameters rapidly reach the surface. The second population consists of globules with a mean diameter of 4 µm, and a third of large globules with diameters ranging from 8 to 12 µm. In colostrum, 70–90% of the fat globules are small but they retain only a small portion of fat. The largest amount of fat is found in the medium-sized globules which represent 10–30% of the total. Approximately 1–4% of the fat is in the remaining 0.01% globules belonging to the population >8 µm in diameter. During the course of lactation, the number of medium and large globules increases, with the mean diameter of the globule population being around 4 µm. This increase in globule size has been reported for several species, including cows [72]. Thus, most of the milk lipids are contained in globules whose diameter is sufficiently large to render them unstable in solution.

In some countries, commercially available milk is routinely homogenized by the dairy industry. In the technology employed, strong pressures are applied to the milk, impelling it through an orifice connected to a decompression chamber. This process involves sudden accelerations and decelerations of the particles in solution, resulting in fat globule rupture with a consequent reduction in size. However, when the size of the fat globules is changed, their protective phospholipid membrane is also ruptured, leading to the loss of natural protection against triglyceride self-digestion by the lipolytic enzymes of milk, the lipases. The effect of homogenization is considered undesirable by the dairy industry, since the production of free fatty acids (FFAs) resulting from triglyceride digestion leads to changes in milk odor and flavor. Pasteurization, a method widely employed in the dairy industry, destroys bacteria as well as enzymes, and is therefore useful to control milk lipolysis.

Human milk possesses at least two different lipases: one present in small amounts whose action is stimulated by plasma, and another which is stimulated by bile salts and whose activity is at least 100-fold that of the first [73], and which has been detected thus far in only gorilla milk in addition to human milk [74]. These enzymes have little activity in milk, either because the stimulus provided by plasma or by bile salts is lacking, or because of the protective membrane surrounding the fat globules [75]. However, as is also the case for cow's milk, they may be activated by the homogenization process.

The technology developed for cow's milk homogenization cannot be applied to human milk. Even the smallest-pressure homogenizers require large volumes of milk and to maintain an appropriate safety margin against contamination, relatively large volumes of approximately 600 ml must be discarded during the process. This virtually rules out the use of this technology for human milk homogenization.

Ultrasonic Homogenization of Human Milk

Since 1986, we have been utilizing a process of human milk homogenization based on ultrasonic vibration. The action of these vibrators is based on the transformation of an electric current of 50/60 Hz into high-frequency energy of 20 Hz. This high-frequency energy is transmitted to a converter where it is transformed into mechanical energy, being then transmitted through a conductor to the solution to be treated. The end of the conductor immersed in the fluid vibrates longitudinally and transfers this movement to the fluid in the form of pressure waves, creating alternating areas of positive and negative pressure. This movement creates countless microscopic bubbles which expand during the negative-pressure phase and collapse during the positive phase. The sudden and violent contraction of the bubbles creates shock waves in the phenomenon referred to as cavitation, which causes the molecules of the fluid to become intensely agitated and which produces a powerful partitioning action at the end of the conductor. The machines can produce emulsions with particles as small as 0.01 μm, homogenize immiscible fluids, accelerate enzymatic or chemical reactions, stimulate bacterial activity, disperse solids in fluids, degas fluids and disintegrate most cells, spores and tissues.

Given the characteristics of the apparatus, it is clear that vibrations of this type applied to human milk may produce undesirable changes in addition to possible beneficial effects such as reduction of the size of fat globules and rupture of bacterial cells. Thus, stimulation of bacterial activity, an increase in auto-lipolysis and changes in the anti-infectious components may be expected to occur in human milk submitted to ultrasonic homogenization.

Some effects of ultrasonic treatment have been analyzed in human milk. It has been demonstrated [68] that fat losses may be reduced from 47.5 to 16.8%

during infusion of bank milk over a 4-hour period if the milk was previously homogenized. Thus, it was concluded that ultrasonic homogenization of expressed human milk [68] can prevent fat separation and loss which otherwise occurs during the tube feeding of infants [63, 67].

However, human milk is a complex fluid, and ultrasonic homogenization might affect labile constituents such as enzymes, immunoprotective substances and bacteria, either as a direct effect of the ultrasonic energy used, or through heat generated during the homogenization process [76, 77]. Therefore, the effects of ultrasonic homogenization on lipolytic activity, immunoglobulins IgA and IgG, lactoferrin, and on bacterial content in expressed human milk were assessed in our laboratory.

Sonication of fresh milk at temperatures $< 45\,°C$ resulted in a significant $(p < 0.01)$ increase of FFA compared to the nonsonicated fresh human milk sample. Addition of bile salts to fresh human milk also resulted in a significant $(p < 0.01)$ increase in FFA production compared to fresh milk, but addition of bile salts to sonicated milk did not produce a further statistically significant increase. As the temperature increased during sonication, the lipolytic activity decreased, approaching zero when the temperature was $> 55\,°C$. Since lipolytic activity was increased by low-temperature ultrasonic homogenization and the addition of bile salts did not lead to a further increase, it seems that the sonication process did not interfere with normal human milk lipase-like activity and that the mechanism whereby sonication increased FFA production was probably similar to that of bile salts. This enhancement of the lipolytic activity of milk by low-temperature sonication might be an advantage for premature infants whose fat digestion may be compromised by bile salt insufficiency [78].

Lipolytic activity was measured in 10 samples each of human milk, either fresh or pasteurized-homogenized, or a 1:1 mixture of both. Despite the low lipolytic activity of fresh milk and pasteurized-homogenized human milk, a mixture of both was able to release significantly more FFAs than either alone $(p < 0.01)$.

Lipolytic activity, with or without added bile salts, decreased as the temperature during homogenization increased and was almost entirely abolished over $55\,°C$. This was probably due to destruction of heat-labile milk lipases [79, 80]. However, even though higher temperatures during homogenization decreased lipolytic activity, this did not interfere with the increased susceptibility of the homogenized fat globules to lipolysis, as shown in the mixing experiment. The lipolytic activity of fresh human milk was low, despite the adequate lipase level, possibly due to the fact that the fat globules were not susceptible to lipase action. In the pasteurized and homogenized human milk, the fat globules were disrupted by homogenization, but the lipolytic activity was low, probably due to the lipase destruction during pasteurization. When mixing the two types of milk, the lipase from the fresh milk could digest the broken fat globules of the homogenized human milk and a greater lipolytic activity was obtained. Ultrasonic homogenization, therefore, appears to make expressed human milk a better substrate for lipolytic digestion.

The mean concentration and range of IgA, IgG and lactoferrin in 16 fresh samples of human milk before and after ultrasonic treatment were also measured and no statistically significant changes were observed when the temperature during sonication was maintained below 45 °C. When temperatures rose above 55 °C, IgA and IgG decreased significantly ($p < 0.01$).

The fact that IgA, IgG and lactoferrin maintained their antigenicity, as measured by Partigen immunoplates, when temperatures during sonication were maintained below 54 °C, implies, but does not prove, that the ultrasonic process did not affect the physiological activity of these constituents. This lack of effect of sonication is desirable.

Since ultrasonication leads to cell disruption, we had surmised that it might destroy bacterial contaminants in milk. We tested this hypothesis against the possibility that the bacterial contaminants are destroyed by the temperature increase occurring in the process. Standardized suspensions of laboratory cultures of *Staphylococcus epidermidis* were added to samples of breast milk immediately before sonication, at initial cell densities of approximately 10^5 organisms/ml. This dense inoculum was chosen to overcome errors due to small quantities of natural flora and to improve the sensitivity for time-kill studies. Viable colony counts were obtained before and after sonication using appropriate agar medium. The solution was sonicated in the temperature range 50–60 °C in 1 °C steps. All bacteria survived when the temperature was maintained below 53 °C, but an abrupt change occurred at 54 °C where survival was 0.05% (n = 6). No organisms survived above this temperature. Sonication failed to kill bacteria in the absence of a temperature rise, and our ultrasonic disrupter could only kill bacteria if the sample temperature rose above 54 °C, at which point temperature-related effects on IgA, IgG and lipolytic activity became apparent. Nevertheless, if bacterial killing is desired, the nature of the ultrasonic process is such that temperatures needed for pasteurization can readily be achieved.

Other long-term effects of human milk homogenization have been investigated. Dhan [81] studied the storage, lyophilization and type of treatment of human milk samples (ultrasonic homogenization, pasteurization and lyophilization) and noted that the recovery of fats and proteins from infused milk after sonication was more than 96% even after 4 months storage. It was concluded that ultrasound homogenization preserves the content of essential fatty acids and a large portion of the long-chain polyenoic fatty acids, and suggested that the increase in FFAs released by ultrasonication may be beneficial for the immature intestine of preterm neonates. Furthermore, the current artificial formulas, unlike human milk (homogenized or otherwise), do not contain docosahexaenoic, arachidonic and other long-chain polyenoic fatty acids. Consequently, the use of artificial formulas could lead to a considerable deterioration of the essential fatty acid status after birth, especially in infants born prematurely. The demand of the fetal and neonatal brain for docosahexaenoic and arachidonic acids increases substantially during the last trimester of pregnancy and the first months of postnatal life.

These results indicated that ultrasonically homogenized human milk should be useful for infant feeding. As well as minimizing fat losses during tube feeding, ultrasonic treatment can increase the digestibility of human milk with regard to lipolysis independently of bile salts. Depending upon the conditions of homogenization, thermolabile substances may be unaffected. Alternatively, bacterial contamination can be eliminated by allowing the temperature to rise during homogenization for pasteurization to take place. The logical outcome of these effects should be the better growth of infants fed homogenized human milk.

On the basis of this rationale, clinical trials to evaluate the usefulness of ultrasonically homogenized human milk in infant feeding were performed at our hospital.

Clinical Trials with Ultrasonically Homogenized Human Milk

Preterm infants were fed banked human milk homogenized by ultrasound, and somatic growth and fat balance were evaluated. To be included in the study, infants had to meet the following criteria: gestational age between 28 and 34 weeks, as determined from the date of the last menstrual period and substantiated by clinical examination of the infant [82]; birth weight below 1,600 g and appropriate for gestational age; absence of congenital anomalies or any major disease, and an ability to begin enteral feeding by the eighth day of life. Informed consent was obtained from the mothers of all infants enrolled in the study. Infants were randomly assigned to one of two groups, receiving either homogenized (group H) or, as the control (group C), nonhomogenized human milk. Any infant withdrawn from the study was replaced by another infant, until each group had 15 infants. Homogenized mild was treated for 15 min with ultrasonic vibration (Tekmar Sonic Disrupter TSD-P 250, Tekmar, Cincinatti, Ohio, USA), not more than 24 h before being offered to the infants. Quality control of homogenization was carried out by microscopic examination of the fat particles [68]. The milk which was offered every 3 h for 30 min was a pool provided by several HIV-donors. Milk was submitted to pasteurization at 62.5 °C for 30 min, and used only after a negative bacteriological test at the milk bank.

Infants were weighed daily. Several other anthropometric measurements were made at the beginning and end of the 20-day study period by a single observer. Crown-to-heel length was assessed using a measuring board with fixed head and side pieces containing a built-in millimeter rule. Skinfold thickness was taken on the left side using a Holtin skinfold calliper with 0.2 mm sensitivity and 10 g/mm² pressure applied for 60 s.

After the 20th day of study, a 72-hour fat balance was carried out on 13 infants from group H and 10 from group C. The quantity of milk consumed by each preterm neonate was estimated by weighing the bottle before and after each feeding. A sample of each lot of milk was stored at $-20\ °C$ for fat determination. After each feeding, the remaining milk was removed from the infusion system, which was washed with a 33% KOH solution to remove all the residue for fat analysis.

Fecal samples were collected by placing the infants on metabolic cradles attached to stainless steel receptacles. The samples were frozen in plastic containers for later fat analysis. The beginning and the end of feces collection were marked by adding carmine to the first and the last feedings.

Sex, gestational age, birth weight, and age distributions at the beginning of the study were similar in both groups. Anthropometric parameters measured at the beginning of the study also showed no statistically significant difference between the groups.

At the 20th day of observation, the group fed homogenized human milk showed better growth than the controls. Weight and triceps skin-fold thickness gain were statistically higher in group H ($p < 0.05$). The gain in length and subscapular skinfold thickness were higher in group H but not statistically so ($p = 0.057$ and 0.061, respectively). As expected, no difference between groups was found in head and thorax circumference.

The daily weight (17 g) and the length (1.6 cm) gain observed in the control group during the study period were similar to those reported in the literature for preterm infants fed banked human milk [38]. The differences between the groups were due to better growth in group H rather than inadequate growth in group C. The better anthropometric measurements shown by group H could be due to greater nutrient intake, better digestibility of the homogenized milk or both. The fat balance was carried out to clarify this point.

The total fat from the milk offered was similar in both groups. The amount of fat left in the infusion system was significantly lower in the group receiving homogenized human milk (5.62 ± 2.39 vs. 11.28 ± 4.43, $p < 0.01$). These losses were due to the fat left in the tubes and some milk residue in the bottle. The milk residue was 81.5 ± 8.4 ml in group C and 80.5 ± 18.5 ml in group H, with a final fat loss due to a residue of 4.51 ± 2.35 g in group C and 2.25 ± 1.53 g in group H. The higher fat loss in group C was not statistically significant. Fat left in the infusion tube was 7.52 ± 3.7 g in group C and 3.37 ± 1.83 g in group H. The fat losses in the infusion systems for both groups were similar to that found after

30 min of human milk infusion in vitro [68]. The total fat losses were 19% of the fat offered in group H and 32% in group C. Fecal fat loss was similar in the two groups.

It was thought that, by breaking up fat globules, homogenization could improve fat digestibility by facilitating the action of lipases [68], but our study showed no improvement in fat digestibility. The milk used was a pool of pasteurized human milk. The pasteurization will have destroyed milk enzymes, including the lipases that play such an important role in milk fat digestion [61]. The shorter the gestational age, the more dependent the infant on human milk enzymes [83]. The mean corrected age for infants during the fat balance study was 38.7 ± 2.2 gestational weeks in group H and 36.6 ± 1.7 weeks in group C. At this age, preterm infants should absorb fat in the same way as term infants, and in fact, for both groups the percentage of fat absorption was similar to that described for the full-term infant fed human milk: 86.6 ± 8.0 and 87.7 ± 6.4 g for groups H and C, respectively [84].

It is worth noting that during the 20 days of receiving ultrasonically homogenized human milk, no infants presented any anomalous clinical symptoms. All infants in both groups remained perfectly well throughout the study.

The better growth of group H raises some doubts about studies comparing the somatic growth of preterm infants fed nonhomogenized human milk and artificial formulas, which are invariably homogenized during manufacture. For mother's milk to be compared to other milks, it is important to ensure that all nutrients in human milk will be completely delivered to the infant.

We then performed studies in which fresh nonpasteurized human milk was fed to younger preterm infants, to analyze the possibility of better fat digestibility after ultrasonic homogenization. A fat balance was recently performed in our hospital on 27 male and female preterm newborn infants of very low weight (1,000–1,500 g). The neonates were divided into three groups: one fed their own mother's nonpasteurized milk, the second pasteurized bank milk, and the third infant formula. The first two groups were further subdivided, the infants acting as their own controls for similar diets. Each infant received either milk homogenized by ultrasound or nonhomogenized milk. The neonates of the two groups participated in two consecutive balances in an alternate random manner, with a 3-day interval between balances. Homogenized milk was given during the first balance followed by nonhomogenized milk during the second and so forth. The third group received milk formula based on powdered cow's milk reconstituted to 8.6% with 5% dextrin-maltose added.

The fat balances, with feces collected for a period of 72 h, showed greater fat absorption in the groups fed homogenized milk from their own mother or the milk bank when compared to nonhomogenized bank milk and infant formula ($p < 0.05$). There was no statistically significant difference in the percentage of fat absorption between infants fed nonsonicated mother's milk and homogenized bank milk. There was also no statistically significant difference between the groups fed nonhomogenized bank milk and infant formula.

Using this methodology, it was possible to demonstrate that preterm infants fed homogenized or nonhomogenized milk from their own mother or homogenized human bank milk have better fat absorption than preterm infants fed nonsonicated human bank milk or infant formula (92, 86 and 87% vs. 78 and 69%, respectively).

Thus, on the basis of the studies reported here, it can be concluded that homogenization of human milk, or human milk formulas, is a safe and easy method for minimizing losses of important nutrients during tube feeding. It is an additional alternative technique which may permit a considerably improved weight gain in preterm neonates.

Human Milk Fortifiers

Much has been done to improve the nutritional quality of human milk. Recent observations have confirmed the beneficial effects of supplementing human milk with protein and minerals [85–87].

Some investigators have recommended the use of fortifiers based on cow's milk, and commercial preparations are available on the market. These fortifiers are sold in small envelopes usually containing 1 g protein and salts which are added to human bank milk or to the mother's own milk. The results obtained with these fortifiers are good, but they have been criticized because they may sensitize the infant. Very early contact of preterm infants with heterologous protein may favor the development of protein allergy, with serious consequences for the child. Although the risk is small, it exists and it should be emphasized.

Another elegant and sophisticated method is to enrich bank milk or the milk of the infant's own mother with a lyophilizate of human milk proteins and with minerals [88–90]. This lactoengineering technique, although very attractive, is mainly limited by the difficulties in preparation, the cost, and the need for large amounts of human milk [91]. Its advantage is that other milk fractions which will become very useful later on, such as IgA and lactoferrin, can be isolated together with the protein [91].

Another sophisticated way of enriching human milk is with casein or lactalbumin hydrolysates [92]. The advantage of hydrolysates is that they supply peptides of high biological value without antigenic characteristics. Protein supplementation has always been of the order of 0.8–1 g/100 ml. Recent studies have demonstrated that growth rate and nitrogen retention improve with the increase in nitrogen fraction, reaching values similar to those of intrauterine gain without a metabolic overload [90].

Human Milk Formula

In our hospital, we work with a population at a very low socioeconomic level. This fact has a series of consequences for preterm infants. The first is that the mothers usually cannot easily reach the hospital. They have many other children at home, may have to work outside the home and, without private transportation, must rely on public services, which are not always efficient. These facts greatly limit the supply of fresh mother's milk to the infant. We have a service of home milk collection, where nurses visit the mothers' homes 3 times a week, and provide the necessary guidance for the mothers. Because of the high risk of contamination of samples collected at home, we opted for milk pasteurization.

However, our milk bank usually has large volumes of mature milk available. Since preterm infants are fed almost exclusively on their own mother's milk or on gestational-age-specific or preterm postpartum-week-specific human milk, mature milk is little used. Significant amounts of milk from donors who delivered term infants are commonly left over and their age is usually 2–4 months after birth.

In our service we have not used mature milk routinely because it has a low protein (0.87 g/100 ml), fat (2.0 g/100 ml) and electrolyte content. Even when this milk is offered in volumes of 200 ml/kg/day, protein calorie and electrolyte levels are not adequate. When these infants are discharged, they will be exposed to a highly hostile environment where their natural defenses will be extremely important.

In view of these considerations, we opted to modify human bank milk on an experimental basis, always trying to preserve its peculiar characteristics but adapting these to preterm feeding. Once adapted, this human milk formula could be used to complement or supplement the mother's own milk.

The method we use is rotary evaporation, concentrating the milk 4-fold. Obviously, all the nutrients in the milk are increased in the same proportion. The only small loss detected during this phase is of fat, which adheres to the evaporation flask, although in amounts that are not sufficient to impair the process.

The basic objective of this milk concentration is to bring lactose, perhaps the only nutrient present in adequate concentration in the original milk, to a concentration above its crystallization point (16 mg/dl at 4 °C). Once precipitated the lactose can be easily removed without using complex ultrafiltration or dialysis methods [93]. Since the lactose concentration in milk is usually 7 mg/dl, 4-fold concentration increases it to levels close to 30 mg/100 ml. The milk is then placed in the refrigerator and later centrifuged to precipitate the excess lactose. With this simple method we can remove approximately 50% of the lactose from concentrated milk (table 1).

After removal of the lactose precipitate, we have a milk in which nutrients are increased 4-fold but with a lactose content only twice the original amount. This

Table 1. Comparison of untreated human bank milk samples and concentrated formulas before and after lactose precipitation

	Human bank milk untreated (n = 32)	Concentrated human bank milk after first lactose precipitation (n = 16)	Concentrated human bank milk after second lactose precipitation (n = 16)
Sodium, mEq/l	12.8 ± 3.2	31.6 ± 20.1	48.3 ± 33.5
Potassium, mEq/l	11.9 ± 1.3	41.0 ± 7.4	48.2 ± 7.8
Calcium, mg%	19.9 ± 4.1	76.1 ± 15.6	67.6 ± 14.8
Phosphorus, mg%	12.3 ± 2.4	36.8 ± 6.6	40.5 ± 7.9
Magnesium, mg%	3.9 ± 0.92	16.1 ± 4.3	16.7 ± 3.6
Protein, g%	0.87 ± 0.17	3.24 ± 0.94	3.66 ± 1.07
Fat, g%	1.94 ± 0.99	7.8 ± 3.8	4.3 ± 2.2
Lactose, g%	7.85 ± 1.94	20.2 ± 4.2	13.2 ± 6.1
Osmolarity, mosm/l	273.6 ± 22.1	708.7 ± 120.2	662.6 ± 65.3
Calories, kcal/100 ml	49.2 ± 8.5	164.0 ± 34.6	105.6 ± 25.3

concentrate is diluted 2-fold, homogenized and pasteurized to be offered later to preterm infants.

As can be seen in table 1, the marked increase in nutrient levels which occurred with the concentration process was not altered by lactose precipitation. In our initial study, the two macronutrients that were a source of concern were lactose and fat. A limitation of the process described above was the slow rate of lactose precipitation, so we tried to accelerate it by freezing followed by centrifugation at 4 °C. When the milk was first centrifuged, only a small amount of lactose was removed. Thus, we started to perform two centrifugations at low temperature. This lowered the level of sugar in the milk considerably, with a consequent reduction in the caloric content of the formula (table 1). To solve this problem, the formula can be enriched with a carbohydrate other than lactose or the milk can be centrifuged for a shorter period so that less sugar precipitates.

The concentration of fat in the original and in the concentrated milk needs to be considered. The starting milk has very low nutrient levels. This fact is not surprising and may be related to the method of expression and to the stage of lactation. This fact has been long denounced by several investigators [63–68] who have emphasized the need for rigid quality control of bank milks not only in terms of bacteriology but also of biochemistry. During the initial study, the appropriate measures to prevent fat loss were not taken, although this could be easily done. It will be very important to take these precautions when performing clinical studies.

When the option is made to dilute the concentrate by half after the precipitation, a formula exclusively based on human milk will be obtained, the composition of which will be much closer to that needed to satisfy the theoretical requirements of preterm infants than the composition of the original milk.

By making the appropriate corrections in terms of lactose and fat losses and by offering this formula to a preterm infant in a volume of 150 ml/kg/day, we shall be satisfying practically all the infant's requirements, except for calcium and phosphorus, which should still be added to the formula. At present we are conducting clinical trials on the use of this human milk formula.

Conclusion

The provision of adequate preterm nutrition requires urgent solutions. In our opinion, all efforts should be directed towards the use of human milk as the basis for feeding the highly susceptible infants. The problems of availability of appropriate amounts of human milk can be easily solved by offering appropriate facilities for milk donation to mothers. The nutritional limitations of human milk can and should be corrected. The technology of the milk industry should be utilized, with enormous advantages to be gained in the scientific alteration and modification of human milk.

Acknowledgements

The authors gratefully acknowledge the following for their support during the tenure of our collaborative research: Hospital das Clinicas of the University of São Paulo Medical School in Ribeirao Preto, São Paulo, Brazil; International Scientific Exchange Program of the Conselho Nacional de Desenvelvimento Scientifico e Tecnologico (CNPq) of Brazil and Natural Sciences and Engineering Research Council (NSERC) of Canada; George Fujisawa Trust Fund, University of British Columbia; Dr. A.G.F. Davidson and staff at the British Columbia Children's Hospital, Vancouver, Canada, and Dr. S. Nakai of the Department of Food Science at the University of British Columbia, Vancouver, Canada.

References

1   Martinez FE, Jorge SM, Gonçalves AL, et al: RN com menos de 1500 g. II: Modificaçoes de conduta e avaliaçao de desempenho des tres epocas distintas nos ultimos 15 anos. J Pediatria 1983;55:113–119.
2   Peacock WG, Hirata T: Outcome in low-birth-weight infants (750 to 1500 grams): A report on

164 cases managed at Children's Hospital San Francisco California. Am J Obstet Gynecol 1981;40:165–172.

3 Cooper PA, Rothberg AD, Davies VA, et al: Comparative growth and biochemical response of very low birth-weight infants fed own mother's milk, a premature infant formula, or one of two standard formulas. J Pediatr Gastroenterol Nutr 1985;4:786–794.

4 Fomon SJ, Ziegler EE, Vasquez HD: Human milk and the small premature infant. Am J Dis Child 1977;131:463–467.

5 Forbes GB: Is human milk the best feed for the low birth weight babies? Pediatr Res 1978;12:434 (422).

6 Heird WC, Anderson TL: Nutritional requirements and methods of feeding low birth weight infants. Curr Probl Pediatr 1977;7:3–10.

7 Lucas A, Goddard P, Baum J: The special care of human milk. Br Med J 1978;2:781–782.

8 Schanler RJ, Garza C, Nichols BL: Fortified mother's milk for very low birth weight infants: Results of growth and nutrient balance studies. J Pediatr 1985;107:437–445.

9 Committee on Nutrition of the American Academy of Pediatrics. Nutritional needs of low-birth-weight infants. Pediatrics 1977;60:519–530.

10 Lubchenko LO, Hansman C, Dressler M, et al.: Intrauterine growth as estimated from live born birth-weight data at 24 to 42 weeks of gestation. Pediatrics 1963;32:793–800.

11 Sala MM, Sala MA: Evoluçao da altura fetal peso do feto da placenta e do indice placentario na segunda metade da gestaçao. Rev Assoc Med Bras 1977;23:88–90.

12 Usher RH, McLean FH: Intrauterine growth of live-born Caucasian infants at sea level: Standards obtained from measurements in 7 dimensions of infants born between 25 and 44 weeks of gestation. J Pediatr 1969;74:901–910.

13 Heim T: Energy and lipid requirements of the fetus and the preterm infant. J Pediatr Gastroenterol Nutr 1983;2(supp 1):S16–S41.

14 Stern L: Early postnatal growth of low birth weight infants: What is optimal? Acta Paediatr Scand Suppl 1982;296:6–13.

15 Forbes GB: Nutritional adequacy of human breast milk for prematurely born infants; in Lebenthal E (ed): Textbook of Gastroenterology and Nutrition. New York, Raven, 1989, pp 27–34.

16 Putet G, Senterre J, Rigo J, et al.: Nutrient balance energy utilization and composition of weight gain in very low birth-weight infants fed pooled human milk or a preterm formula. J Pediatr 1984;105:79–85.

17 Reichman BL, Chessex P, Putet G, et al: Partition of energy metabolism and energy cost of growth in the very low birth-weight infant. Pediatrics 1982;69:446–451.

18 Ziegler EE, O'Donell AM, Nelson SE, et al: Body composition of the reference fetus. Growth 1976;40:329–341.

19 Hanning RM, Zlotkin SH: Amino acid and protein needs of the neonate: Effects of excess and deficiency. Sem Perinatol 1989;13:131–141.

20 Räihä NCR: Milk protein quantity and quality and protein requirements during development. Adv Pediatr 1989;36:347–368.

21 Fomon SJ, Filer LJ, Anderson TA: Recommendations for feeding normal infants. Pediatrics 1979;63:52–59.

22 Poskitt EME: Infant feeding: A review. Hum Nutr 1983;37A:271–286.

23 von Sydow G: A study of the development of rickets in premature infants. Acta Paediatr Scand 1946;33(suppl 12).

24 Anderson GH, Atkinson S, Bryan MH: Energy and macronutrient content of human milk during early lactation from mothers giving birth prematurely and at term. Am J Clin Nutr 1981;34:258–265.

25 Atkinson SA, Anderson GH, Bryan MH: Human milk: Comparison of the nitrogen composition in milk from mothers of premature and full-term infants. Am J Clin Nutr 1980;33:811–815.

26 Atkinson SA, Bryan MH, Anderson GH: Human milk: Difference in nitrogen concentration in milk from mothers of term and preterm infants. J Pediatr 1987;93:67–69.

27 Butte NF, Garza C, Johnson CA, et al.: Longitudinal changes in milk composition of mothers delivering preterm and term infants. Early Hum Dev 1984;9:153–162.

28   Gross SJ, David RJ, Bauman L, et al.: Nutritional composition of milk produced by mothers delivering preterm. J Pediatr 1980;96:641–644.
29   Lemons JA, Moye L, Hall D, et al.: Differences in the composition of preterm and term human milk during early lactation. Pediatr Res 1982;26:113–117.
30   Moran JF, Vaughan R, Stroop S, et al.: Concentrations and total daily output of micro-nutrients in breast milk of mothers delivering preterm: A longitudinal study. J Pediatr Gastroenterol Nutr 1983;2:629–634.
31   Schanler RJ, Oh W: Composition of breast milk obtained from mothers of premature infants as compared to breast milk obtained from donors. J Pediatr 1980;96:679–681.
32   Ferlin ML, Santoro JR, Jorge SM, et al.: Total nitrogen and electrolyte levels in colostrum and transition human milk. J Perinat Med 1986;14:251–257.
33   Lucas A, Gore SM, Cole TJ, et al.: Multicentre trial on feeding low birth-weight infants: Effects of diet on early growth. Arch Dis Child 1984;59:722–730.
34   Brooke OG, Wood C, Barley J: Energy balance, nitrogen balance and growth in preterm infants fed expressed breast milk, a premature infant formula and two low-solute adapted formulas. Arch Dis Child 1982;57:898–904.
35   Cooper PA, Rotheberg AD, Pettifor JM, et al.: Growth and biochemical response of premature infants fed pooled preterm milk or special formula. J Pediatr Gastroenterol Nutr 1984;3:749–754.
36   Davis DP: Adequacy of expressed breast milk for early growth of preterm infants. Arch Dis Child 1977;52:296–301.
37   Gaull GE, Rassin DK, Raiha NCR: Protein intake of premature infants: A reply. J Pediatr 1977;90:507–510.
38   Gross SJ: Growth and biochemical response of preterm infants fed human milk of modified infant formula. N Engl J Med 1983;308:237–241.
39   Palhares DB, Jorge SM, Martinez FE: Avaliaçao antropométrica de recém-nascidos pré-termo alimentados com leite humano do banco de leite ou com fórmula industrializada de leite de vaca. J Pediatria 1987;63:129–132.
40   Raiha NCR, Heinonen K, Rassin DK, et al.: Milk protein quantity and quality in low birth-weight infants. I: Metabolic responses and effects on growth. Pediatrics 1976;57:659–674.
41   Roberts SB, Lucas A: The effects of two extremes of dietary intake on protein accretion in preterm infants. Early Hum Dev 1985;12:301–307.
42   Tyson JE, Lasky RE, Mize CE, et al.: Growth, metabolic response and development in very low birth-weight infants fed banked human milk or enriched formula. I: Neonatal findings. J Pediatr 1983;103:95–104.
43   Atkinson SA, Radde IC, Anderson GH: Macromineral balances in premature infants fed their own mother's milk or formula. J Pediatr 1983;102:99–106.
44   Cooke RJ, Nichoalds G: Nutrient retention in preterm infants fed standard infant formulas. J Pediatr 1986;108:448–451.
45   Anderson GH, Bryan MH: Is the premature infant's own mother's milk the best? J Pediatr Gastroenterol Nutr 1982;1:157–159.
46   Brooke OG: Nutrition in the preterm infant. Lancet 1983;i:514–515.
47   Nutrition Committee Canadian Paediatric Society: Feeding the low birth-weight infant. Can Med Assoc J 1981;124:463–467.
48   Juto P: Human milk stimulates B cell function. Arch Dis Child 1985;60:610–613.
49   Narayanan I: Feeding of preterm infants with expressed human milk. Indian Diary 1985;37:97–100.
50   Narayanan I, Murthy NS, Prakash K, et al.: Randomised controlled trial of effect of raw and holder pasteurised human milk and of formula supplements on incidence of neonatal infection. Lancet 1984;ii:1111–1113.
51   Welsh JK, May JT: Anti-infective properties of breast milk. Pediatrics 1979;94:1–9.
52   Hamosh M: Bile-salt-stimulated lipase of human milk and fat digestion in the preterm infant. J Pediatr Gastroenterol Nutr 1983;2(suppl 1):S248–S251.
53   Berseth CL, Lichtenberger LM, Morriss FH Jr: Comparison of the gastrointestinal growth-promoting effects of rat colostrum and mature milk in newborn rats in vivo. Am J Clin Nutr 1983;37:52–60.

54   Haasis P, Cutrell P, Epplein D, et al.: Nursing care; in Goldsmith JP, Karotkin EH (eds): Assisted Ventilation of the Neonate. London, Saunders, 1981, pp 81–97.

55   Hess JH, Lundeen EC: The Premature Infant, ed 2. Philadelphia, Lippincott, 1949, pp 119–121.

56   Jorges JE, Magnusson JH, Wrelind A: Casein hydrolysate: A supplementary food for premature infants. Lancet 1946;ii:228–232.

57   Jarvenpaa A, Raiha NCR, Rassin DK, et al.: Preterm infants fed human milk attain intrauterine weight gain. Acta Pediatr Scand 1983;72:239–243.

58   Lewis MA, Smith BA: High volume milk feeds for preterm infants. Arch Dis Child 1984;59:779–781.

59   Herbst JJ: Development of suck and swallow. J Pediatr Gastroenterol Nutr 1983;2(suppl 1): S131–S135.

60   Nutrition Committee Canadian Paediatric Society: Statement on human milk banking. Can Med Assoc J 1985;132:750–752.

61   Clark RM, Hundrieser KH, Ross S, et al.: Effect of temperature and length of storage on serum-stimulated and serum-independent lipolytic activities in human milk. J Pediatr Gastroenterol Nutr 1984;3:567–570.

62   Hamosh M, Berkow S, Bitman J, et al.: Handling and storage of human milk specimens for research. J Pediatr Gastroenterol Nutr 1984;3:284–289.

63   Brooke OG, Barley J: Loss of energy during continuous infusions of breast milk. Arch Dis Child 1978;53:322–325.

64   Greer FR, McCormick A, Loker J: Changes in fat concentration of human milk during delivery by intermittent bolus and continuous mechanical pump infusion. J Pediatr 1984;105:745–749.

65   Naratanan I, Sing B, Harvey D: Fat loss during feeding of human milk. Arch Dis Child 1984; 59:475–477.

66   Spencer SA, Hull D: Fat content of expressed breast milk: A case for quality control. Br Med J 1981;282:99–100.

67   Stocks RJ, Davies DP, Allen F, et al.: Loss of breast milk nutrients during tube feeding. Arch Dis Child 1985;60:164–166.

68   Martinez FE, Desai ID, Davidson AGF, et al.: Ultrasonic homogenization of expressed human milk to prevent fat loss during tube feeding. J Pediatr Gastroenterol Nutr 1987;6:593–597.

69   Fransson GB, Lonnerdal B: Iron copper zinc calcium and magnesium in human milk fat. Am J Clin Nutr 1984;39:185–189.

70   King RL, Luick JR, Litman II, et al.: Distribution of natural and added copper and iron in milk. J Dairy Sci 1959;42:780–790.

71   Richardson T, Guss PS: Lipids and metals in fat globule membrane fractions. J Dairy Sci 1965;48:523–530.

72   Ruegg M, Blanc B: The fat globule size distribution in human milk. Biochim Biophys Acta 1981;666:7–14.

73   Hernell O, Olivecrona T: Human milk lipases. I: Serum stimulated lipase. J Lipid Res 1974;15: 367–374.

74   Freudenberg E: A lipase in the milk of the gorilla. Experientia 1966;22:317.

75   Patton SJ, Keenan TW: The milk fat globule membrane. Biochim Biophys Acta 1975;415:273–309.

76   Jelliffe DB: Unique properties of human milk: Remarks on some recent developments. J Reprod Med 1975;14:133–137.

77   Widdowson EM: Protective properties of human milk and the effects of processing on them. Arch Dis Child 1978;53:684–686.

78   Lebenthal E, Lee PC: Alternate pathways of digestion and absorption in early infancy. J Pediatr Gastroenterol Nutr 1984;3:1–3.

79   Heitlinger LA: Enzymes in mother's milk and their possible role in digestion. J Pediatr Gastroenterol Nutr 1983;2(suppl 1):S113–S119.

80   Williamson S, Finucane E, Ellis H, et al.: Effect of the heat treatment on human milk absorption of nitrogen, fat sodium, calcium and phosphorus by preterm infants. Arch Dis Child 1978; 53:555–563.

81   Dhar J: Effect of Ultrasonication, Lyophilization, Freezing and Storage on Lipids and Immune

Components of Human Milk; Master's thesis, University of British Columbia, Vancouver, 1988, p 109.

82   Dubowitz LMS, Dubowitz V, Goldberg C: Clinical assessment of gestational age in newborn infant. J Pediatr 1970;77:1–10.

83   Jensen RG, Clark RM, deJong FA, et al.: The lipolytic triad: Human lingual, breast milk, and pancreatic lipases. Physiological implications of their characteristics in digestion of dietary fats. J Pediatr Gastroenterol Nutr 1982;1:243–255.

84   Fomon SJ, Ziegler EE, Thomas LN, et al.: Excretion of fat by normal full-term infants fed various milks and formulas. Am J Clin Nutr 1970;23:1299–1313.

85   Schanler RJ: Human milk for preterm infants: Nutritional and immune factors. Semin Perinatol 1989;13:69–77.

86   Modanlou HD, Lim MO, Hansen JW, et al.: Growth, biochemical status, and mineral metabolism in very low birth-weight infants receiving fortified preterm milk. J Pediatr Gastroenterol Nutr 1986;5:762–767.

87   Salle B, Senterre J, Putet G, et al.: Effects of calcium and phosphorus supplementation on calcium retention and fat absorption in preterm infants fed pooled human milk. J Pediatr Gastroenterol Nutr 1986;5:638–642.

88   Ronnholm KAR, Simell O, Siimes MA: Human milk protein and medium-chain triglyceride oil supplementation of human milk: Plasma amino acids in very low birth-weight infants. Pediatrics 1984;74:792–799.

89   Svenningsen NW, Lindroth M, Lindquist B: A comparative study of varying protein intake in low birth-weight infant feeding. Acta Paediatr Scand Suppl 1982;296:28–31.

90   Salle B, Putet G, Senterre J, et al.: Doit-on toujours alimenter les enfants de très faible poids de naissance avec un lait de lactarium? Arch Fr Pédiatr 1987;44:157–159.

91   Voyer M, Seterre J, Rigo J, et al: Human milk lacto-engineering: Growth, nitrogen metabolism and energy balance in preterm infants. Acta Paediatr Scand 1984;73:302–306.

92   Tonz O, Schubiger G: Feeding of very low birth-weight infants with breast milk enriched by energy, nitrogen and minerals. Helv Paediatr Acta 1985;40:235–247.

93   Ronnholm KAR, Sipila I, Siimes MA: Human milk protein supplementation for the prevention of hypoproteinemia without metabolic imbalance in breast milk-fed very low birth-weight infants. J Pediatr 1982;101:243–247.

Dr. F.E. Martinez, Department of Pediatrics, University of São Paulo Medical School, 14048 Ribeirao Preto, São Paulo (Brazil)

Simopoulos AP, Dutra de Oliveira JE, Desai ID (eds): Behavioral and Metabolic
Aspects of Breastfeeding. World Rev Nutr Diet. Basel, Karger, 1995, vol 78, pp 74–113

....

Breastfeeding in Australia

Margaret Lund-Adams, Peter Heywood[1]

Nutrition Program, University of Queensland, Brisbane, Australia

Contents

[1] The authors would like to thank the Nursing Mothers' Association of Australia and
the Lactation Resource Center for the valuable assistance they have given in the collection
of information and resource materials.

Introduction

In this paper we review published literature to assess the overall breastfeeding situation in Australia[2]. Three separate, but related, topics are discussed at length: trends in breastfeeding since 1940, factors affecting breastfeeding, and efforts to support and promote breastfeeding. The paper concludes with a synthesis of this information and recommendations for future action.

Breastfeeding Trends

In this section we describe the general infant feeding situation in Australia for two periods: before 1940, and from 1940 to the present. For the latter period, we review data on breastfeeding rates available from both health services and special surveys.

Before 1940

Aborigines

Prior to the arrival of Europeans, it is believed that breastfeeding of Aboriginal infants was universal, a necessary component of survival. Mothers would carry their babies while gathering food, allowing continual suckling. If the

---

[2] For those not familiar with Australia, the country has a national, federal government, as well as government at the state and territory level. New South Wales, Queensland, South Australia, Tasmania, Victoria and Western Australia make up the six states. The two territories are Australian Capital Territory and Northern Territory.

need arose, other lactating women would substitute for the mother to ensure that breast milk was available for the infant. Reports suggest that breast milk was the main food well into the second year of life. As the Aborigines were not herdsmen, animal milk was not part of the weaning diet [1].

With the spread of European settlement, the traditional hunting and gathering life-style of the Aborigines was restricted. Over time, Aborigines drifted towards the towns and some settled on the missions [1]. With the increase of urbanization and contact with Europeans, breastfeeding prevalence decreased [2]. The place of residence and the degree of assimilation into the mainstream Australian life-style affects Aboriginal breastfeeding patterns. In remote areas where the traditional, tribal ties are stronger breastfeeding prevalence remains higher [1, 3].

Europeans

In 1788, the first European settlers came to Australia bringing with them the contemporary ideas of their homeland. The infant feeding methods advocated at the time in the British Isles, in order of preference, were breast milk from the mother, breast milk from a wet nurse, and hand feeding with either animal milks, pap or panada (mixtures of cereal, such as bread, rice, or barley, and fluid, such as water, broth, or milk – panada was of a thicker consistency than pap) [4, 5].

Early colonial records make little mention of infant feeding practices. However, it is known that the British government endorsed breastfeeding and was concerned about possible infant deaths on the long voyage to Australia. Women were prevented from embarking until their babies were weaned and no baby was allowed to be weaned before 6 months of age. Even after that age, a convict woman had to get medical permission to allow her to wean her infant [4, 5].

During the early years of settlement, food rations were limited. Livestock was scarce and the availability of milk and milk products was extremely limited. Hence, convict mothers had little choice but to breastfeed. However, with time, food supplies improved. Fresh fruit and vegetables became available, as did milk from cows and goats. Modified cow's milk became a common breast milk substitute. Apart from the nutritional composition of cow's milk, the type of preparation techniques and the utensils used to feed infants posed serious health problems. Artificially fed infants had high mortality rates [4–7].

In addition to convicts, by 1810, some free settlers had arrived. Their numbers increased significantly from the 1820s, as the availability of land grants was recognized. With the arrival of free settlers came the demand for wet nurses. The 'well-to-do' mothers could afford to pay for the services of lactating convict women. Some foundling institutions also paid breastfeeding convict mothers to feed orphaned infants. However, the employment of wet nurses encouraged the practice of baby farming. The babies of the wet nurses were 'farmed out' and many died as a result [4–6].

Female Factories were used to house convict women and their children before they were assigned to free settlers as servants. In a review of the Female Factory at Hobart in 1841, it was found that the infants were generally weaned between 9 and 12 months and then separated from their mothers [4–6]. Gandevia [6] suggests that most women probably weaned their infants onto a pap or panada of cereal, perhaps boiled with some salt meat and then tied in a rag for the child to suck.

Data recorded in 1851 suggest that babies cared for in convict nurseries, where breastfeeding was the rule, had lower mortality rates than those in the outside community where early weaning was common. The difference in mortality was particularly marked in infants < 1 month of age. It was recorded in the 1899 Australasian Medical Gazette that a minimum of 52% of infants under the age of 6 months farmed out with other families died. In foundling homes where artificial feeding was practiced, the mortality rate was even higher. In institutions where infants were breastfed, the mortality rates were lower [6].

Condensed and dried milks became available in Europe during the 1850s. It was not long before they reached Australia. As the quality of fresh milk was variable and generally poor, condensed milks became very popular, keeping well in the warm Australian climate. By 1883, 27 brands of infant food were available on the Australian market. The first commercially prepared infant formula was introduced into Australia in the early 1930s. Despite considerable advances in the formulation of breast milk substitutes, mortality rates and the occurrence of rickets and scurvy remained higher in artificially fed infants well into the 20th century. Summer diarrhea was a great hazard for artificially fed infants, and gastroenteritis remained the major cause of infant death until about 1950 [5–9].

Despite the availability of processed milk products and infant formula, at the end of the 1800s and in the early 1900s, the available information suggests that most Australian mothers still breastfed their infants. During this time, it seems that the literature advising mothers on infant feeding practices encouraged breastfeeding [4, 8, 9]. Some promoted breastfeeding as a moral duty to the baby and accused bottle feeding mothers of being irresponsible. A woman's role became centered on child care as motherhood was praised as a service to the new nation and child-raising techniques were seen as a predominant influence on child health [9, 10]. It was believed that breastfeeding was the first secret in the prevention of infant deaths. The second secret was the provision of clean food, and weaning in the summer months was discouraged [6].

After 1940

Data are not available to document precisely the decline in breastfeeding rates. However, a number of Tasmanian reports and surveys are available, the earliest of which is the 1924 Hobart Child Welfare Association Report. These documents suggest that most infants were still breastfed between 1924 and 1940. However, the prevalence of breastfeeding declined dramatically between 1941 and 1952 [7, 11, 12]. Data collected routinely in Victorian clinics around the same time also showed a similar decline in breastfeeding rates [13]. This decline continued through the 1950s and 1960s, throughout Australia, as artificial feeding continued to grow in popularity [8, 14].

In addition to the increased availability and affordability of artificial feeds, Smibert [15] suggests that part of this change resulted from the move from home to hospital delivery. The practices in many hospitals were not conducive to breastfeeding. He concluded that the routines and rules imposed on new lactating mothers were introduced mainly by male obstetricians and unmarried nurses who knew little about the practicalities of breastfeeding. The attitude of medical staff towards breastfeeding was discouraging and often misinformed. Mobbs and Mobbs [16] report that in the early 1970s, maternity hospitals were display cases for the promotion of infant formulas: free samples, displays, leaflets, and advertisement-laden nursing magazines.

During the late 1960s and 1970s, women became increasingly critical of the medicalization of childbirth, and women's groups pressed for changes to allow more 'natural' childbirth and 'natural' infant feeding practices [17]. Calls were made for better breastfeeding education and more positive attitudes towards it from doctors, nurses, hospital administrators, and departments of health [18, 19].

During the 1980s, response to these calls resulted in changes in maternity hospital policies and the attitudes of health professionals. Today, breastfeeding is encouraged and it is expected that new mothers will breastfeed rather than bottle feed. It is now generally the norm for maternity hospitals to encourage immediate breastfeeding and demand feeding, allow rooming in, and not routinely give bottle complements [13, 17].

However, Minchin [20] warns that there is still considerable work to be done. She advocates that further changes in the political, health, and social arenas are still necessary to overcome the barriers to maintaining longer breastfeeding duration. She sees a need for improved pregnancy counselling and hospital practice, improved maternity benefits, workplace childcare with provision for breastfeeding, reeducation of health professionals at various levels, public support for breastfeeding women, and planning health services to support breastfeeding.

More specific details of these general trends in breastfeeding since 1940 are available from routinely collected data and surveys at the national, state, and local

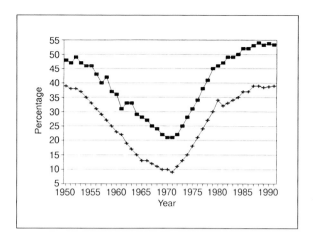

Fig. 1. Infants fully breastfed at Infant Welfare Centers in Victoria between 1950 and 1991 [22]. ■ = 3 months; + = 6 months.

levels. Breastfeeding prevalence rates at hospital discharge are equated with levels of breastfeeding initiation, and rates at later ages, such as 3, 6 or 12 months, indicate duration.

Routinely Collected Data

Since 1944, statistics have been collected on very large numbers of infants (up to 70,000 annually) by all Infant Welfare Centers throughout Victoria. When statistics were first kept, 55% of babies were fully breastfed for at least 3 months and about 42% for 6 months. Over the ensuing years, a decline continued until 1971, the nadir, when only 21% of mothers were fully breastfeeding their 3-month-old infants and 9% were fully breastfeeding their infants at 6 months [8, 13, 21, 22]. As can be seen in figure 1, in 1972, the trend to bottle feed was reversed by a percentage point or two. Breastfeeding rates climbed quickly, and by the early 1980s, prevalence rates were similar to 1950 levels.

Criticism has been voiced concerning use of Infant Welfare Center data to monitor trends in the total population, as attending mothers are self-selecting and motivated. However, it does permit trends to be monitored over time amongst clinic attenders and Scott [23] claims that 96% of all babies born in Victoria attend a center.

Similar to the Victorian picture, South Australian annual data, also collected through clinics, showed a dramatic return to breastfeeding between 1972 and 1980 [24].

Table 1. National breastfeeding surveys

Year	Location	Sample size	Sampling method	Study design	Breastfeeding prevalence, %					Reference
					at HD	6 weeks	3 months	6 months	12 months	
1983	Australia	60 hospitals	representative on the basis of number of deliveries and sociodemographic profile of population serviced	C-S	85[a]	NA	NA	NA	NA	25
1983	Australia	unknown	available data from state and territory departments of health: routinely collected data and surveys	NA	NA	72	54–55	40–42	10–12	25
1989–1990	Australia	3,520 women aged 18–50 years, with a child/ children aged ≤5 years	stratified multistage area sample of households which ensured persons within each state or territory had a known, and in the main, an equal chance of selection	C-S	77[b]	NA	NA	NA	NA	27

HD = Hospital discharge; C-S = cross-sectional; NA = not available.
[a] Exclusively breastfed (no complementary milk feeds).
[b] Percentage of mothers who have ever breastfed.

National Surveys

1983 National Breastfeeding Survey. The first national survey was conducted by the Commonwealth Department of Health in 1983 in response to the need for data by the Australian delegation to the 1983 World Health Assembly [25].

Questionnaires were sent to 60 hospitals throughout Australia sampled to represent major maternity hospitals, private nursing homes, regional hospitals, district base hospitals, and smaller country hospitals. Responses to this survey indicated that 85% of mothers were fully breastfeeding their babies at hospital discharge. In Western Australia and Northern Territory the figure was as high as 95–97%. The data were not analyzed to give prevalence rates by socioeconomic or ethnic group [25].

Obtaining national data on the prevalence of breastfeeding at later infant ages proved more difficult. Palmer [25] gathered information from existing administrative statistics of state and territory health departments, and from available survey data. These data were originally collected for different purposes and used different methodologies, definitions, infant age groups, and reporting periods [25]. Hence, collating the varied information was not easy and certainly could not give a truly representative picture. The results are displayed in table 1.

The national survey indicated that despite the high rate of breastfeeding initiation, there was a drop by 6 weeks of age. This decline continued with approximately 54% of 3-month-old infants and 40% of 6-month-old infants receiving breast milk. By 12 months of age, only 10% were breastfed. Palmer [25] concluded that encouraging mothers to breastfeed longer needed to become a major health target.

1989–1990 National Health Survey. The 1989–1990 National Health Survey was the first in a new series of 5-yearly surveys to be conducted by the Australian Bureau of Statistics (ABS). It was carried out throughout the 12-month period October 1989–September 1990. Households were selected randomly using a stratified multi-stage area sample which ensured that persons within each state and territory had a known and, generally, equal chance of selection. In total, 22,000 dwellings were selected, covering 57,000 persons (about 1 in 300 of the Australian population) [26, 27].

Trained ABS interviewers conducted personal interviews with adult residents of selected dwellings. Women, aged 18–64 years, were invited to complete an additional written questionnaire relating to women's health concerns. In addition to English, questionnaires were available in 11 other languages to prevent exclusion of non-English speakers [26]. The last 3 of the 16 women's health questions concerned breastfeeding. Only women aged 18–50 years with children aged ≤5 years were asked to complete them.

Table 2. State breastfeeding surveys

Year	Location	Sample size	Sampling method	Study design	Breastfeeding prevalence, %					Reference
					at HD	6 weeks	3 months	6 months	12 months	
1983	South Australia	17,266	all child health centers	L	78	NA	41	30	NA	24
1984	South Australia	1,140	all child health centers	L	82	NA	53[a]	37[b]	NA	24
1985	South Australia	811	all child health centers	L	80	NA	52[a]	36[b]	NA	24
1986	South Australia	808	all child health centers	L	80	NA	50[a]	36[b]	NA	24
1952	Tasmania	945	NA	NA	62[c]	NA	44[c]	NA	NA	11
1969	Tasmania	547	all child health centers	L	53[c]	NA	17[c]	NA	NA	11
1974	Tasmania	396	all child health centers	L	57[c]	NA	23[c]	NA	NA	12
1979	Tasmania	460	all child health centers	L	66[c]	NA	41[c]	NA	NA	28
1984–1985	Tasmania	460	all child health centers	L	77[c]	64[c]	51[c]	39[c]	12[c]	29, 30
1979	Western Australia	686	available records from 18 child health centers: 12 metropolitan and 6 country	L	82	74	64	43	9	31
1984–1985	Western Australia	911	stratified, two-stage sample of 24 child health centers: 16 metropolitan and 8 rural	L	84[c]	66[c]	53[c]	41[c]	11[c]	29, 30
					86	74	62	45	13	

HD = Hospital discharge; L = longitudinal; NA = not available.
[a] At 4 months of age.
[b] At 7 months of age.
[c] Exclusively breastfed (no complementary milk feeds).

The results presently published give, by age of mother, the number of women who have 'ever breastfed'. Seventy-seven percent of all women surveyed reported having breastfed (table 1). Across the age groups of 18–24, 25–34, 35–44, and 45–50 years, the proportion of women who had ever breastfed was 77, 79, 70, and 36%, respectively [27].

Figures on the duration of breastfeeding by age of mother have also been published. Of those babies 'ever breastfed', 23% were breastfed for <3 months. Forty-seven percent were breastfed for <6 months and 79% for <12 months. Only 7% were breastfed for ⩾18 months [27]. The data do not give the prevalence of breastfeeding at particular infant ages.

The ABS data need to be interpreted with caution. Duration of breastfeeding usually refers to the length of time an infant was breastfed, the infant being completely weaned. However, the published ABS data do not refer only to weaned children. The length of time of breastfeeding of current breastfeeders was mixed with that from previous breastfeeders. At the time of interview, a woman may have been breastfeeding for only 2 months but may have continued to breastfeed for another 10 months. However, the duration of her breastfeeding is recorded and analyzed as only 2 months. Thus the proportions calculated were distorted. Another concern is that women who have more than one child aged ⩽5 years have their breastfeeding histories counted more than once. This group of women may have particular breastfeeding patterns that could distort the calculated proportions. In addition to these problems, the definition of breastfeeding was not clear. Hence, the percentages of children who were exclusively breastfed or partially breastfed cannot be calculated.

State Surveys

Data collected from various statewide surveys are shown in table 2. It is interesting to note that all these surveys used longitudinal data collected by child health centers. Thus, the samples are likely to be biased toward self-selected motivated women and limits their value in estimating prevalence rates in the population generally. However, it does permit trends to be monitored over time. The trends seen in the states of South Australia, Tasmania, and Western Australia are similar to those previously discussed from Victoria.

As discussed earlier, data collected annually from South Australian Mothers' and Babies' Health Association Clinics showed a dramatic increase in breastfeeding prevalence rates between 1972 and 1980. When surveys on clinic attenders were done between 1983 and 1986, this rate of increase had slowed [24].

Data from Tasmania in 1952, 1969, 1974, 1979, and 1984–1985 showed exclusive breastfeeding prevalence rates at hospital discharge and at 3 months [11, 12, 28–30]. Both rates declined, reaching the lowest point in 1969, and then began to climb. By 1984–1985, exclusive breastfeeding prevalence was 77 and 51% at hospital discharge and 3 months, respectively.

The 1951 Western Australian Department of Public Health Report [8] noted that the number of mothers breastfeeding their infants had decreased to 42% at 6 months. A subsequent report for 1970 [8] noted that the average duration of lactation decreased from 97 days in 1960, to 75 days in 1964, and to 73 days in 1968. As is shown in table 2, by 1979 the prevalence of breastfeeding at hospital discharge had risen to 82% and was back up to 43% at 6 months [31]. In 1984–1985, the rates were slightly higher than those reported in 1979 [29].

Local Surveys

Table 3 summarizes the result of fifteen studies conducted throughout the six Australian states from 1963 to 1984. Despite some variation, the overall trend seems consistent with that recorded in infant health centers already presented. Both breastfeeding initiation and prevalence rates during the first year of life seem to have declined until the early 1970s and then climbed.

New South Wales. The four New South Wales studies were all carried out in Sydney between 1972 and 1984 [32–36]. The lowest prevalence rates were recorded in the northern suburbs in 1972. By 1977, in a similar sample of mothers from the same area, these prevalence rates had dramatically increased. In just 5 years, the prevalence rates at hospital discharge, 6 weeks, 3 and 6 months had increased by 13, 30, 39, and 36%, respectively [32]. However, it must be noted that residents of northern Sydney suburbs tend to be of higher socioeconomic status and this may have influenced the dramatic increases and the high prevalence rates. Lower breastfeeding rates were found in the same year, 1977, among women giving birth in the obstetric hospitals in the metropolitan area of Sydney and at the Royal Hospital for Women [33–35]. Similar hospital discharge rates of breast-feeding to these were reported in the Western Metropolitan Health Region (76%), 7 years later, in 1984. However, a higher proportion of socioeconomically disadvantaged people live in the Western Metropolitan Health Region [36]. These Sydney studies suggest that in areas of higher socioeconomic status, breastfeeding prevalence rates increased more quickly and reached higher levels.

Queensland. One of the Queensland studies was conducted in an isolated rural area. Records of all births in the Cunnamulla hospital during 1974 and 1975 were reviewed. Caucasian women of low socioeconomic status and Aboriginal women had very low breastfeeding initiation rates: 15 and 16%, respectively. In contrast, 63% of Caucasian women of higher socioeconomic status were discharged from hospital breastfeeding. The author concluded that socioeconomic status was more important than race in predicting breastfeeding initiation [37].

Two other Queensland studies were conducted in the metropolitan area of Brisbane and Ipswich between 1972 and 1984. The authors used available records from selected Maternal and Infant Health Clinics, the inherent selection bias again limiting generalization of the findings to the whole population [38, 39]. However, because similarly selected infant records from the same four clinics were reviewed for 1972 and 1982, it does allow comparisons over time in these clinics' population. During this 10-year period, there was a dramatic increase in breastfeeding prevalence rates at 3, 6, and 12 months. Rates rose from 31, 14, and 0.4% to 63, 48 and 17%, respectively [38]. The 1984 study was prospective at four different clinics, chosen to cover the range of social classes. As the same four clinics were not used in 1982 and 1984 and the prevalence rates for 1984 are reported for the age groups 0–3, 3–6, and 9–12 months, the data are not directly comparable. However, the much higher prevalence rates in 1984 do suggest that breastfeeding popularity was still on the increase among clinic attenders [38, 39].

In addition to clinic data, between the mid-1970s and mid-1980s, data from annual birth records at the Brisbane Mater Mothers' Hospital also show increases in breastfeeding rates. At hospital discharge, breastfeeding rates increased from 46 to 75% in public (noninsured) women and from 59 to 89% in private (insured) women between 1975 and 1987 [40].

South Australia. Hankin [41] in her 1965 Adelaide study estimated that 80, 29, and 16% of mothers were breastfeeding their infants at hospital discharge, 3, and 6 months, respectively. Ten years later, Boulton and Coote [42, 43] reported that 73% of infants left hospital breastfed. However, by 3 and 6 months of age, this proportion was 39 and 31%, respectively.

Tasmania. Apart from the statewide surveys conducted in Tasmania, one was carried out in the capital, Hobart, in 1964. No details about the study sampling methods or design are available to us. It reported a low prevalence of exclusive breastfeeding at hospital discharge (52%) and at 3 months (19%) [11].

Victoria. None of the studies conducted in Melbourne suburbs between 1963 and 1982 are representative of the total Melbourne population [44–46]. Records from five Infant Welfare Centers, selected to reflect a range of socioeconomic status, showed a slight drop in exclusive breastfeeding at 3 months of age from 32 to 28% between 1963 and 1964 [44]. The other two studies, which will be referred to below, were conducted in areas of low socioeconomic status with either high proportions of single mothers, migrants, refugees and/or non-English speakers [45, 46].

Western Australia. In 1979, a prospective study of the growth and feeding practices of 205 infants born in Perth was initiated. These infants, selected randomly from the Midwife Record of Birth Data collected by the Western Australian Public Health Department, were at least second generation born in Australia, non-Aboriginal, from two-parent families, full-term, singletons, and weighed $\geqslant 2{,}500$ g at birth. The subjects were mainly from middle-class families; over 70% were from the middle two groups of a four-point social scale. The results showed that not only did a high proportion of women commence breastfeeding but a considerable proportion continued throughout the first year. At 6 weeks, 3,

Table 3. Local breastfeeding surveys

Year	Location	Sample size	Sampling method	Study design	Breastfeeding prevalence, %						Reference
					at HD	6 weeks	3 months	6 months	12 months		
1972	northern suburbs of Sydney, NSW	218	baby health records of a similar sample of mothers to the 174 who delivered in 1977	L	73	50	31	12	NA	32
1977	northern suburbs of Sydney, NSW	174	random selection of mothers delivering in June 1977 in every hospital in the northern suburbs of Sydney and those who travelled to deliver in an inner-city midwifery hospital	L	86	80	70	48	NA	32
1976–1977	Sydney, NSW	250	representative sample of infants selected from obstetric hospitals in the Sydney metropolitan area using a cluster sample technique	C-S	79	NA	31	NA	NA	33, 34
1977	Sydney, NSW	410	all mothers who delivered at the Royal Hospital for Women, Sydney, over a 2-month period	L	77	59	41	22	NA	35

Year	Location	n	Description	Study type						Ref
1984	Western Metropolitan Health Region, Sydney, NSW	8 hospitals (158 mothers)	all public hospitals in the Western Metropolitan Health Region with > 1,000 births annually	C-S	76[a]	NA	NA	NA	NA	36
1974–1975	Cunnamulla, Queensland	103 in total:	all births in Cunnamulla Hospital, 1974 and 1975, categorized as follows:	C-S						37
		40	Aborigines		15	NA		NA	NA	
		25	Caucasians of low socioeconomic status		16	NA		NA	NA	
		38	Caucasians of high socioeconomic status		63	NA	NA		NA	
1972	Brisbane and Ipswich, Queensland	229	random selection from infants who had attended, for at least 12 months, one of four maternal and infant health clinics selected from a working-class area (n = 1), upper-middle-class suburb (n = 1), and socially mixed areas (n = 2)	L	NA	NA	31	14	0.4	38

Table 3. (continued)

Year	Location	Sample size	Sampling method	Study design	Breastfeeding prevalence, %					Reference
					at HD	6 weeks	3 months	6 months	12 months	
1982	Brisbane and Ipswich, Queensland	272	as above	L	NA	NA	63	48	17	38
1984	Brisbane and Ipswich, Queensland	143	infants born in 1984 who attended maternal and infant health clinics in four centers (chosen to cover the range of social classes)	L	NA	NA	83[b]	65[b]	25[b]	39
1965	Adelaide, South Australia	108	NA	L	80	54	29	16	0.9	41
1975–1977	Adelaide, South Australia	391	systematic sampling (every 4th baby born) of babies born between November 1975 and July 1976 at Queen Victoria Hospital	L	73	NA	39	31	8	42, 43
1964	Hobart, Tasmania	200	NA	NA	52[c]	NA	19[c]	NA	NA	11
1963	5 suburbs of Melbourne, Victoria	820	records from 5 Infant Welfare Centers to reflect a range of socioeconomic status	C-S	NA	NA	32[c]	NA	NA	44

Year	Place	N	Selection	Type						Ref
1964	as above	815	as above	C-S	NA	NA	28[c]	NA	NA	44
1979	Brunswick, Melbourne, Victoria	304	all consecutive births in Brunswick municipality	L	82	44	29	16	8[d]	45
1982	3 inner city suburbs, Melbourne, Victoria	220	NA	C-S	79	NA	57	NA	NA	46
1979–1980	Perth, Western Australia	205	random selection from midwife record of births for 1979	L	NA	83	77	64	25	47–49
1980–1982	Perth, Western Australia	48	Vietnamese infants born between 1980 and 1981, selected from three sources: those living in a migrant hostel and attending an antenatal clinic, through an informal community network, and midwife record of birth data	L	81	NA	56	43	23	50

HD = Hospital discharge; NSW = New South Wales; L = longitudinal; NA = not available; C-S = cross-sectional.

[a] Breastfeeding prevalence amongst hospitalized women ⩾ 3 days postpartum.
[b] Prevalence figures are for infants in age groups 0–3, 3–6, and 9–12 months.
[c] Exclusively breastfed.
[d] At 44 weeks.

6, and 12 months, the following proportions of women continued to breastfeed: 83, 77, 64, and 25%, respectively [47–49]. Around the same time in Perth, Reynolds et al. [50] also collected prospective data on 48 Vietnamese infants. At 3 and 6 months of age, their breastfeeding prevalence rates were considerably lower (56 and 43%, respectively). However, by 12 months the prevalence rates in the two groups were similar [47–50].

Methodology Problems

The lack of clarity and consistency of breastfeeding definitions has caused problems in data interpretation and comparisons across studies [51–53]. In the surveys already discussed and summarized in tables 1–3, some reported only exclusive breastfeeding, some reported both exclusive and partial breastfeeding, while others gave no definition. The Australian Institute of Health, which makes recommendations for the review of health goals and targets, has stressed the need for clarifying the definition of breastfeeding and how it relates to mixed breast and bottle feeding [54].

The distinctions between feeding groups and clear definitions must be decided in relation to the nature and purpose of the study. The scheme and framework developed by the Interagency Group for Action on Breastfeeding forms a solid basis on which to assist researchers and agencies in this area [52].

Infant feeding practices change frequently during the early months. A baby may be exclusively breastfed, mixed fed and then artificially fed all in the first month of life. Hence, researchers need to focus on short time periods or specific ages [53]. Among the studies reviewed, most selected to measure breastfeeding prevalence rates at specific ages. However, rarely was the description clear how a child's age was defined. Does breastfeeding at 6 months mean the infant was breastfed at some stage during the 6th month, for the entire 6th month, or between the 22nd and 26th week? [Department of Health, Housing and Community Services (DHHCS): Review of the Implementation in Australia of the WHO International Code of Marketing of Breast-Milk Substitutes; Unpubl. report, 1992]. The lack of clear age definition compounds the difficulty of cross-study comparisons. One study grouped infants into 3-month age categories, such as 0–3, and 3–6 months.

Recall bias is a problem in many surveys. Winikoff [53] cautions that the passage of time may cause mothers to forget certain events, or heap certain events into convenient ages (i.e. full months or half years) or perceived acceptable times. Even though the results of one Australian study by Eaton-Evans and Dugdale [55] indicated that the duration of breastfeeding was generally remembered accurately, it is likely that the details of changes in feeding practices, particularly in the early months when many changes may occur, are frequently forgotton or recalled inaccurately.

Caution needs to be taken when reviewing hospital discharge breastfeeding rates. The time in hospital is too short to determine whether lactation has been successfully established. However, it does measure initiation and give an indication of those mothers who have 'ever breastfed' and, therefore, is useful [56]. Presently, individual hospitals collect this information, but it is not readily available on a regional, statewide or national basis.

Sample selection has posed a problem in many of the studies outlined in tables 1–3, making comparisons between studies and generalization of results to the total population difficult. Many studies used child health clinic data. The samples are inherently biased, as the groups are self-selecting. Even if a high proportion of all mothers attend clinics in the early months of their infant's life, by 12 months of age this proportion is greatly reduced. Hence, longitudinal studies over 12 months which rely on clinic attenders are certainly not representative of the general population. As already stated, they are, however, useful to monitor changes over time among attenders at the same clinic. Other sampling problems included convenience samples, unclear descriptions of sampling methods and study design, and small sample sizes.

Results from studies conducted in specific geographical areas were often confounded by socioeconomic status, education level, race, and migrant status. These variables need to be controlled, if correct comparisons of breastfeeding prevalence rates between geographical areas are to be made.

Summary of Breastfeeding Trends

Prior to the arrival of Europeans, breastfeeding among Aboriginal infants was probably universal. With the spread of European settlement, the life-style of Aborigines changed dramatically and breastfeeding prevalence decreased among those who moved into towns and cities. In remote areas, where traditional ties are stronger, the prevalence rates remain higher.

Among the first European settlers, infant feeding tended to follow patterns of the time in the British Isles. Infant mortality was high, particularly among those who were not breastfed. Overall breastfeeding initiation and duration remained high until approximately 1940. By 1950, there had been a significant decline in breastfeeding duration and this trend continued through to the early 1970s when as few as 20–25% of infants were fully breastfed at 3 months of age.

During the late 1960s and early 1970s, women became increasingly critical of the medicalization of childbirth, and calls were made for changes in hospital procedures and the attitudes of the medical staff in an effort to promote breastfeeding. At the same time, overall breastfeeding initiation and duration began to increase and continued to do so until the mid-1980s when the rate of increase slowed. By the mid-1980s fully breastfed prevalence rates at 3 and 6 months of age had reached approximately 50 and 35%, respectively, similar to those seen in the mid-1950s.

Factors Affecting Breastfeeding

Evidence on the factors affecting breastfeeding is reviewed in this section with particular emphasis on socioeconomic status and ethnicity. Information on the reasons for stopping breastfeeding is also reviewed.

Socioeconomic Status

A case-control study of breast cancer, diet and breastfeeding history in Brisbane, Queensland, between 1981 and 1985 provides evidence of the effect of education on infant feeding practices [57]. The responses of the 844 randomly selected controls supplied information on the feeding practices of first-born infants from the 1920s to the 1970s. There was a steady decline in both the proportion of infants breastfed and the duration of lactation from the 1930s to the 1970s. However, these trends were abruptly reversed in the early 1970s when breastfeeding popularity rose. Better-educated women were the early adopters and leaders of the infant feeding fashion. They led the way in abandoning breastfeeding and turning to bottle feeding in the late 1950s. Better-educated women also led the return to breastfeeding in the early 1970s [Siskind et al., unpubl. report, 1991].

A survey in Victoria in 1963, using Infant Welfare Clinic records, found the following results. Breastfeeding rates at 3 months of age were lowest in the area where there was a high proportion of low-income families. A higher proportion, 35%, of the women who delivered in private hospitals were totally breastfeeding at 3 months compared with 25% from public hospitals. A higher proportion of wives of professional men breastfed their infants than did women from other groups in the community [44].

Further evidence of the social gradient of breastfeeding rates was provided in the 1970s by Borda et al. [35] in New South Wales, Cox [37] in Queensland, Boulton and Flavel [58] in South Australia, and Smibert [21], and Williams and Carmichael [45] in Victoria. In 1975, Boulton and Flavel [58] found that socioeconomic status, indirectly measured by a mother being a private or public hospital patient, was positively associated with a higher incidence of breastfeeding.

At the time of hospital discharge, 49% of public mothers and 76% of private mothers were breastfeeding. Smibert states 'there is no doubt that the trend away from the bottle has occurred mainly in the more educated classes'. In 1978, he surveyed patients discharged from private, intermediate, and public wards of the Royal Women's Hospital in Melbourne. The percentages of babies with a birthweight > 3,000 g who were fully breastfed on discharge were 83, 79, and 66%, respectively. Assuming that the type of hospital ward gives an indication of current economic status and education level, these figures suggest that the return to breastfeeding first occurred in the higher socioeconomic groups [21]. Williams and Carmichael [45] in a study in a poor municipality in Melbourne in 1979 also found successful breastfeeding was positively correlated with better education. Mothers with ≥ 11 years of schooling breastfed their babies longer than less-educated mothers.

As discussed previously, annual birth records from the Mater Mothers' Hospital in Queensland showed increases in breastfeeding rates at discharge between 1975 and 1987. Despite this increase being experienced by both public and private patients, the private patients maintained higher rates. At hospital discharge, breastfeeding rates increased from 46 to 75% in the public patients and from 59 to 89% in the private patients between 1975 and 1987 [40].

During the 1980s, other studies continued to reveal more evidence of the positive association between socioeconomic status and breastfeeding. A study in 1982 in inner-city Melbourne clearly showed that with increasing occupational rank, mothers were more likely to initiate breastfeeding and to continue breast-feeding for longer. Single, unsupported mothers had even lower rates of breast-feeding initiation and duration than other women of low occupational rank [46]. In the 5-year longitudinal study in Perth [47–49], breastfeeding was more common and continued for a longer duration in the upper two social ranks. This was found again, in 1984–1985, in both Western Australia [29, 31], and Tasmania [29]. Not only were more mothers from the higher social ranks breastfeeding on discharge from hospital, but they were breastfeeding for longer. In a study on bottle feeding mothers in Sydney, 82% had at some stage attempted breastfeed-ing. By the time their infants were 4 weeks old, 64% of mothers from the working-class area had ceased breastfeeding, compared to only 31% of mothers from the affluent area. It is of concern that those with lower incomes are the ones choosing the more expensive feeding option. Not only did the more affluent women breastfeed for longer, they also sought professional advice on feeding more often [59].

Manderson [17] also found that middle-class mothers who face feeding difficulties are more likely to access available assistance than working-class women. Working-class women tend to lack the confidence to ask for help and are intimidated by institutions, including voluntary groups of middle-class women.

They struggle with their breastfeeding problems alone rather than call for assistance.

A New South Wales study published in 1992 suggests the continuing existence of the differential between socioeconomic groups. When their infants were 4 months of age, women who had not completed the Higher School Certificate were 17-fold more likely to have stopped breastfeeding than women with tertiary education [60].

The increasing prevalence of breastfeeding in Australia since the early 1970s has occurred largely in the absence of government interventions. The avant-garde in the return to breastfeeding has been the mothers from higher socioeconomic groups [8]. They have led the move and societal change has reinforced it and promoted change in other groups. The following factors have been suggested as contributors: growth of the womens' movement; establishment of woman-to-woman support groups such as the Nursing Mothers' Association of Australia (NMAA); movement of consumers towards 'natural is best'; increased emphasis on the importance of bonding of the mother and child dyad, and increasing knowledge of the benefits of breastfeeding [14, 61].

Ethnicity

Breastfeeding prevalence varies across different ethnic groups. In this section we review studies looking at breastfeeding among just two ethnic groups: Australian Aborigines, and Vietnamese immigrants. It must be noted that these studies are not recent, and that patterns may have changed in the last decade.

Aborigines

Table 4 summarizes the results of breastfeeding studies in four Aboriginal groups [62–65]. Even though the Queensland study [63] relied on clinic records, data were available on >90% of the children born in Cherbourg Aboriginal Settlement in the 5 selected study years. The data show that at 3 months of age, breastfeeding prevalence was highest in 1953 (40%), fell from then until the late 1960s (13%), and had begun to rise by 1972 (30%). This pattern is similar to that seen in the non-Aboriginal Australian population.

A study in Western Australia in 1981 [65] found differences among Aboriginal groups from various locations. Groups in remote areas seemed to have continued the traditional infant feeding practice of breastfeeding for as long as possible. At 12 months of age, 96% of infants in the remote sample were still being breastfed. With increasing urbanization, the prevalence of breastfeeding at all ages decreased. Gracey et al. [65] associate this decline with the loss of traditional culture and increasing affluence.

Cox [37] (table 3) found that there was no difference in obstetric outcomes and breastfeeding rates between Aboriginal and low-socioeconomic-status Anglo-Australian women. He concludes that ultimately class, not race, 'is of more importance in predicting the major antenatal and delivery characteristics of women in Western Queensland' [p. 23].

Vietnamese Immigrants

Research among Vietnamese immigrant women in four Australian cities between 1977 and 1983 indicated that breastfeeding was generally an unpopular feeding choice [14, 50, 66–69].

Traditionally, Vietnamese infants were breastfed for approximately 18 months. However, it was not uncommon for children to be breastfed into their 4th year [66]. More recently, many urban Vietnamese women, even in Vietnam, have adopted a negative attitude towards breastfeeding. They see it as shameful and unwesternized. The growing trend towards bottle feeding seen in urban Vietnam seems to be accelerated among women who migrate to Australia. Breastfeeding, in addition to being shameful and unwesternized, is seen as an interference with outside employment, a survival necessity for the newly immigrated family [17, 67, 68]. Matthews and Manderson [66] found that return or entry into the work force was the major reason given by respondents for not breastfeeding in Vietnam (30%) and Australia (73%). In addition, inadequate milk supply was also a common reason given for bottle feeding when in Vietnam and in Australia.

Vietnamese and other migrant women, due to English language problems and lack of knowledge about the availability of services, tend not to call for help when feeding problems are encountered [17].

Reasons Given for Stopping Breastfeeding

As already outlined, breastfeeding initiation in Australia is high but there is a marked reduction in the prevalence over the first 3 months. Some report that the first week following hospital discharge is the critical period when support is most needed [DHHCS, unpubl. report, 1992]. The most common reason given by mothers in Australia since the 1950s for stopping breastfeeding is that of 'inadequate milk supply' [25, 28, 34, 35, 44, 46, 56, 59, 60, 70, 71]. This response may result from a lack of understanding about proper positioning and the mechanisms of supply and demand in breastfeeding. Mothers often interpret their baby's demand for extra feeds, crying or sleeping problems as indicators of insufficient milk supply [35, 56, 70, 71]. Others have judged inadequacy on the basis of weight gains. Smaller weight gains of breastfed infants compared to artificially fed infants and the deviation from the ideal curve set by artificially fed

Table 4. Aboriginal breastfeeding surveys

Year	Location	Sample size	Sampling method	Study design	Breastfeeding prevalence, %					Reference
					at HD	6 weeks	3 months	6 months	12 months	
1978	Rural area, New South Wales	146 mothers or guardians of children aged 1 month to 5 years	NA	C-S	52	NA	31	16	NA	62
1953	Cherbourg Aboriginal Settlement, Queensland	65	data gathered from all available Infant Welfare Clinic records	L	NA	NA	40	8	0	63
1958	as above	57	as above	L	NA	NA	23	5	0	63
1963	as above	47	as above	L	NA	NA	13	6	0	63
1968	as above	52	as above	L	NA	NA	13	4	0	63
1972	as above	47	as above	L	NA	NA	30	13	0	63
1980–1981	Perth, Western Australia	127 maternal/infant pairs	birth cohort included consecutive births of Aboriginal infants at metropolitan Perth hospitals between November 1980 and July 1981, who continued to reside in the city for at least 6 months	L	82	NA	50	NA	19	64

1981	Western Australia	645 subjects in total:	convenience sample from 3 locations:	C-S						65
			(1) remote tribal or mission-oriented communities	196	100	100	100	100	96	
			(2) partly urbanized, largely nontribal, living in or outside big country towns	376	NA	NA	71	66	61	
			(3) nontribal, living in towns located in rural districts of the southwest	73	NA	NA	NA	<50	33	

HD = Hospital discharge; C-S = cross-sectional; NA = not available; L = longitudinal.

infants can cause mothers concern [19, 35, 44, 56, 72]. Palmer [25] comments that this perceived insufficiency of milk is 'probably related to poor management of early problems, unavailability of appropriate support (social and medical) in the early weeks after parturition and inappropriate preparation'.

Other authors echo Palmer's suggestion that lack of support by health professionals, family, and friends contributes to early breastfeeding cessation [56, 70, 71, 73, 74]. Minchin [74] suggests that most early breastfeeding problems have their origins in poor positioning at the breast. She argues that mothers need to be given appropriate advice early by skilled health professionals who understand the process of breastfeeding and who are convinced of its importance. Mothers need encouragement to develop confidence in their ability to breastfeed and to seek help for breastfeeding problems, such as sore nipples, early. Support systems, such as the NMAA group, lactation consultants and child health services, are available but many women who need them do not use them. Lower usage rates have been reported among lower socioeconomic groups and immigrant women. The suggested reasons for not using support services vary: English language difficulties, not knowing these services exist or how to acquire their assistance, lack of confidence, and intimidation by institutions [17, 73]. Bailey and Sherriff [70, see also ref. 71] report that mothers from lower socioeconomic groups generally do know about these support systems, but choose not to use them.

The issue of women's employment has also been discussed as another possible factor which affects breastfeeding practices [17, 19, 44, 56, 61, 66–68, 75]. In Australia, women are entering and remaining in the work force in increasing numbers. By 1986, 48% of women ≥15 years were in the labor force. The increase in participation of women has been particularly large among married women aged 25–54 years, the time traditionally associated with child rearing. In 1986, 63% of married women aged 20–24 years and 59% of married women aged 25–54 years were in the work force. Despite this increase, many married women work part-time. In 1986, of those married women employed, 45% worked <35 h per week [76].

Since 1979, maternity leave benefits have been widely available for Australian women. Under federal and state awards, women are entitled to 52 weeks of unpaid maternity leave. In some areas of the public sector, paid maternity leave may also be available, usually 12 weeks. To be eligible for maternity leave, employees must have been in continuous service for at least 12 months and must be full-time or part-time employees, not casual or seasonal workers. A study, conducted by the Australian Institute of Family Studies, has looked at issues surrounding maternity leave in Australia. All women who gave birth to a child during one week in May 1984 were surveyed 18 months later. A response rate of 50% resulted in 2,012 responses [77].

The study showed that the major exit from the labor force by women occurs after the birth of the first child. Of those employed during pregnancy, 60%

returned to the work force, mainly for financial reasons. The main reason given for not returning by the remaining 40% was that they did not want to work while their child was young. The next most common reason, given by about half of the respondents, was breastfeeding. Many women cited the desire to breastfeed in conjunction with not intending to work while the child was young [77].

Bundrock [56] believes that many mothers are not aware that it is possible to breastfeed and be employed simultaneously. Even so, it is difficult to do. Maternity leave provisions in Australia do not conform with those of the International Labour Organisation Convention which state that women should be provided with breastfeeding breaks when they return to work [77]. The National Health and Medical Research Council (NHMRC) [75] and others have noted that supportive work site health promotion policies are needed. These policies need to consider flexible work times and breastfeeding breaks, childcare facilities, provision of facilities to express and store breast milk, education, and group support [20, 56, 61, 75].

Despite the difficulties, women rarely report work as the main reason for stopping breastfeeding. The highest reported percentage of responses that mentioned work as a reason to stop breastfeeding was merely 3% [19, 28, 34, 35, 44, 70]. However, work may be a reason why women do not commence breastfeeding. As already noted, Vietnamese women are unwilling to breastfeed because they see it as an interference with work or the possibility of obtaining a job [17, 67, 68]. Matthews and Manderson [66] found that 73% of the Vietnamese women they interviewed reported that return or entry into the work force was the major reason for not breastfeeding. In 1977, when Borda et al. [35] interviewed 410 mothers who delivered at the Royal Hospital for Women in Sydney, 85 chose to bottle feed. 'Returning to work' was given as the reason for bottle feeding by 9% of the women. Among primiparas, this percentage rose to 21%.

Summary of Factors Affecting Breastfeeding

Socioeconomic status, including education as well as income, is the most potent factor affecting breastfeeding. Women of high socioeconomic status led the move away from breastfeeding during the 1950s and 1960s. However, they also led the reversal in this trend, in both initiation and duration, in the 1970s and 1980s.

Among Aborigines, those on settlements, in towns and in urban areas have tended to show overall rates similar to those seen in the non-Aboriginal population, and in these settings there is evidence that socioeconomic status may be a more important influence than ethnicity per se. In remote locations, breastfeeding has remained more prevalent than in urban areas.

Breastfeeding rates among Vietnamese women, one of the more recent immigrant groups in Australia, are comparatively low.

The most common reason for stopping breastfeeding is 'inadequate milk supply'. Most studies have concluded that this response is a result of a lack of understanding about breastfeeding, the relative isolation of new mothers, and the lack of professional and social support. The large number of married women in the work force points to the continuing need for employers to provide appropriate facilities and support.

Support for Breastfeeding

In this section we review international initiatives and the efforts of Australian government and nongovernment organizations to support and promote breast-feeding.

International Initiatives

In 1979, a joint WHO/UNICEF Meeting on Infant and Young Child Feeding was held. Due to concern over increasing rates of artificial feeding and associated health problems, the meeting recommended the support of breastfeed-ing and the development of an international marketing code for infant formula and weaning foods [78]. An 'International Code of Marketing of Breast-Milk Substitutes' (WHO Code) was later drafted and adopted at the 34th session of the World Health Assembly in May 1981 [79]. Throughout the 1980s, UNICEF continued to promote breastfeeding as part of its GOBI startegy. In 1989, WHO and UNICEF released a statement on 'Protecting, Promoting and Supporting Breastfeeding: The Special Role of Maternity Services'. In this document, ten steps to successful breastfeeding are outlined [80]. In 1990, the Convention on the Rights of the Child, the World Summit for Children, and the Innocenti Declara-tion all again called international attention to the importance of breastfeeding. In 1991, WHO and UNICEF launched the Baby Friendly Hospital Initiative to promote the adoption of the ten steps to successful breastfeeding in maternity facilities throughout the world. Also in 1991, the World Alliance for Breastfeeding Action (WABA) was formed [81–83].

Government Support

National Goals and Targets
In 1976, the NHMRC of Australia released a statement on 'Feeding of Infants and Young Children'. It was amended in 1980 and endorsed breastfeeding as the most suitable method of feeding Australian infants [84]. 'Increase breast-

feeding' was the first of eight Australian Dietary Goals adopted in 1979. To facilitate public education, this goal was later translated into the dietary guideline message: 'promote breast-feeding' [85]. The Dietary Guidelines for Australians have recently been reviewed and in June 1992 the breastfeeding guideline was changed to 'encourage and support breastfeeding' and lowered from 1st to 8th position [75].

The Better Health Commission Taskforce, in 1986, and the Health Targets and Implementation (Health for All) Committee, in 1988, both included a breastfeeding target when setting nutrition targets for improving the health of Australians. By the year 2000, the aim is for 95% of Australian infants to be breastfed at discharge from hospital and at least 80% to be breastfed at 3 months of age [85, 86]. The primary intention of the target is to increase the duration of breastfeeding, not merely increase initiation rates [54].

Monitoring National Goals and Targets

Although Australia has quantifiable national breastfeeding targets, until 1989 there was no system to collect data or monitor trends over time to assess progress towards these national goals. As mentioned earlier, for the first time, the ABS collected breastfeeding information in its 1989–1990 National Health Survey. The Australian Institute of Health [54, 87] looks to these planned 5-yearly ABS surveys to allow the monitoring of progress towards the achievement of the national breastfeeding target. However, due to problems discussed earlier, modification of the ABS questionnaire will be needed to improve the future usefulness of the data collected.

Despite clinic nurses recording the feeding status of the individual infants they see, not all health departments, on a regional, territory, or state basis, aggregate and publish these collected statistics. Victoria and South Australia give examples of how breastfeeding data can be collected through the state clinic system. It would be useful for all infant health services to commence, reinstate, or continue collection and collation of breastfeeding data. However, with national direction, collection methods would need to be standardized to ensure comparability of data across regions, states, and territories.

International Code of Marketing of Breast-Milk Substitutes

Australia was among the 118 member states to vote in favor of the WHO Code at the World Health Assembly in May, 1981 [79, 88]. Subsequently, a number of government initiatives have been taken, although Australia is one of many countries which has not yet fully adopted the WHO Code.

In October 1981, the NHMRC released a statement in support of the WHO Code [84]. One of the first government initiatives was to negotiate with Australian manufacturers and importers of infant formula. In May 1983, an 'Industry Code of Practice for the Marketing of Infant Formulas', in which the industry voluntarily controls its marketing practices by self-regulation, was agreed to by both industry and the government [88]. This was later updated in 1986, and is often referred to as the 'Australian Agreement' [DHHCS, unpubl. report, 1992].

In 1993, the NHMRC set up a working party on the implementation of the WHO Code. In June 1984, the NHMRC released WHO Code implementation guidelines and the working party published its report in March 1985 [88].

Responsibility for implementation of the WHO Code was given mainly to state and territory health authorities. Recommendations covered the areas of hospital care and discharge, pre- and postnatal services, community education, education of health professionals, community support groups, monitoring of nutritional status, educational materials, code monitoring, and dealing with code breaches [88].

Additionally, between 1984 and 1986, the NHMRC also released statements on complementary feeding in newborn nurseries, marketing of unconventional foods for infants, and guidelines on the storage and pooling of human milk [84].

In September 1991, a joint statement by the Minister for Health, Housing and Community Services and the Minister for Justice and Consumer Affairs was issued regarding the desirability of breastfeeding, its 'public good' and indicating continuing government support for the objectives of the WHO Code. A working party including government, industry, retail and consumer representatives was set up in October 1991 with the view of developing a document similar to the WHO Code particularly for Australia which would be capable of authorization by the Trade Practices Commission. The working party is moving towards the development of a series of agreements [89]. The first of these, 'Marketing in Australia of Infant Formulas: Manufacturers and Importers' was signed by industry and government in May 1992. This agreement supersedes the previous voluntary industry agreement and prohibits the advertising of infant formula to the public. The Australian government is now in the process of setting up an independent panel which will monitor the implementation of the agreement. This should ensure improved accountability and enforcement. Future agreements for authorization by the Trade Practices Commission are planned to delineate the responsibilities of retailers and pharmacies, and bottle and teat manufactures and importers [89, 90]. This series of codes will be combined with a set of government policies directed at the health sector [DHHCS, unpubl. report, 1992].

Monitoring of the WHO Code

Until now, there has been no formal national government monitoring of the WHO Code. State and territory health departments, to whom this responsibility was originally given, tend to be divided into geographic regions where services and policies are often determined locally by the regional director. Generally, it seems that high priority has not been given to monitoring WHO Code violations. However, in some areas, state health departments do record reported breaches of the code on an informal basis. Hence, what does occur is ineffective, as there is no mechanism for collating information regionally, statewide, or nationally through the health sector [DHHCS, unpubl. report, 1992].

Unofficial monitoring of the WHO Code is done by the NMAA, the Australian Lactation Consultants Association (ALCA), the Australian Consumers' Association and other concerned members of the public. Since 1987, the NMAA has recorded reported violations to the WHO Code. Several hundred reports suggest that inappropriate practices still do occur within the health system. Wide regional variation in violations indicate that the standards of practice are not consistent and some locations need particular improvement [56]. In 1990, a new coalition of concerned groups was formed: the Australian Coalition for Optimal Infant Feeding (ACOIF). The coalition aims to create a national structure that will enable reporting and assessment of the breast milk substitute situation [91].

Compliance with the WHO Code

It is difficult to measure and monitor compliance with the WHO Code.

In New South Wales, in October 1984, an infant-feeding survey was conducted in all eight public hospitals in the Western Metropolitan Health Region with > 1,000 deliveries annually [36]. One of the main aims of the study was to assess compliance with the recently released NHMRC guidelines on implementing the WHO Code in hospitals [36, 88]. The results showed that, in principle, the policies of all hospitals encouraged breastfeeding as the preferred method of feeding, and recent changes in some hospital practices were evidence of supportive action. Of the hospitals surveyed, 63% encouraged mothers to breastfeed their infants immediately or within an hour of birth, 50% permitted 24-hour rooming in, 100% avoided routine complementary feeds and 100% had nursing personnel assisting with breastfeeding. Areas highlighted as still needing improvement included allowing frequent flexible feeding, avoidance of prelacteal feeds, and discontinuing the giving of samples of formula at discharge [36].

In 1991, a national review of the implementation of the WHO Code [92] was conducted [DHHCS, unpubl. report, 1992]. Time limitations did not allow observation of practices in a representative sample of Australian institutions. Therefore, visits were made to six selected locations, four urban and two rural. In each of these, a hospital, two baby health centers, two supermarkets and two pharmacies were visited. In addition to these visits, questionnaires were mailed across Australia to all maternity hospitals with at least 500 births annually, all medical schools, all state health departments, professional and mothers' support groups, manufacturers and distributors of infant formulas, and chain pharmacies and supermarkets [92]. Advertising in professional journals and lay magazines between 1971 and 1991 was also reviewed [DHHCS, unpubl. report, 1992].

It should be noted that at the time of the 1991 review, the new 'Marketing in Australia of Infant Formulas: Manufacturers and Importers' agreement had not been signed, but the self-regulating industry code, Australian Agreement, was in operation. The main findings of this review are reported in the following paragraphs [DHHCS, unpubl. report, 1992].

(1) At present there is no formal system to monitor the implementation of the WHO Code. Informal monitoring, mainly by nongovernment organizations, has occurred but has been ineffective as no enforcement mechanism exists.

(2) The implementation of the WHO Code is complicated by the division of responsibilities for food legislation, fair trading, and health care provision across national, territory, and state boundaries.

(3) The legislative role of the national government is confined to import and export controls over labelling of milk and milk products. Under their respective Food Acts, territory and state governments are responsible for safety, quality, and labelling of infant formulas. The state standards for infant formula are in line with the codex standard for infant formulas and follow the labelling requirements set by WHO. At both national and state levels, there are laws governing misleading or deceptive advertising or conduct in the course of trade or commerce.

(4) In addition to the problems of differences across states, many state health departments have recently undergone regionalization, moving much of the policy and service delivery responsibility from the center to the region. However, from the survey conducted, all state health departments reported having policies which prohibit promotion of formulas in hospitals and baby health centers [92].

(5) Putting the baby to the breast soon after delivery, rooming in, demand feeding, and the avoidance of prelacteal and complementary feeding were reported in all hospitals visited. Mothers received support and encouragement from hospital staff to breastfeed.

(6) Education materials for parents, distributed at the hospitals and health centers visited, did not fully meet the WHO guidelines. Most material neglected to warn of the difficulty of reversing a decision not to breastfeed, of the social and financial implications of using formula, and of the health hazards associated with inappropriate foods or feeding methods.

(7) All baby health nurses interviewed knew about the WHO Code. However, only 43% of other health professionals interviewed knew about it. Generally, among all health professionals, knowledge about the Code contents is limited.

(8) The amount of formal training given to medical students about infant feeding is limited.

(9) Formula manufacturers seem to have complied generally with the self-regulatory Code. Advertising of infant formula to the general public rarely occurs and labelling meets requirements. However, infant formula has been advertised extensively in professional journals.

(10) Price promotion of formulas, bottles, and teats by pharmacies occurs frequently [92].

(11) Advertising and promotion of bottles, teats, and non-formula breast milk substitutes for infants under 4–6 months of age is not restricted in Australia and occurs widely [92].

(12) There is no national system for collecting prevalence data of breastfeeding at particular infant ages. It appears that the incidence of initial breastfeeding is high, but that the rates drop quickly after discharge from hospital. Adequate support for breastfeeding mothers in the first week following discharge is critical.

(13) Emerging social pressures for women to return to work soon after delivery may affect breastfeeding rates.

World Summit and Baby Friendly Hospital Initiative

The World Summit for Children agreed upon a set of goals for the survival, protection, and development of children. This 'World Declaration and Plan of Action' was formally ratified by the Australian government in May 1991. As part of the response, it has been suggested that the Baby Friendly Hospital Initiative will be incorporated into federal government plans for program implementation. The Minister for Health, Housing and Community Services has written to all state and territory health ministers to inform them of the Baby Friendly Hospital Initiative and urge them to adopt it [93]. Information has also been sent to the Australian Hospitals Association suggesting its incorporation into health promotion initiatives [UNICEF Australia: Baby Friendly Hospital Initiative: Preliminary Meeting; unpubl. report, 1992].

UNICEF Australia convened a meeting in April 1992 to further discuss the implementation of the Baby Friendly Hospital Initiative in Australia. The meeting indicated widespread support for nationwide implementation. A task force is being established to explore the possibility of incorporating the Baby Friendly Hospital Initiative goals into hospital accreditation procedures [94] and hospital staff education programs [UNICEF Australia: Baby Friendly Hospital Initiative: Preliminary Meeting; unpubl. report, 1992].

Nongovernment organizations play a vital role in the encouragement and support of breastfeeding in Australia. In addition to working with individual mothers offering advice and support, they have actively lobbied for the full implementation and monitoring of the WHO Code, kept records of code violations, raised issues about maternity hospital practices, and conducted public education.

The NMAA was founded in Melbourne in 1964 with the aim of encouraging, educating, and supporting mothers who wished to breastfeed. The original group of six mothers has grown to one of the largest womens' groups in Australia with 14,500 members in 1991. Cumulative membership over the first 27 years reached 112,000 women. In addition to members, thousands of other women each year are assisted by the 1,700 trained breastfeeding counsellors who are available seven days a week for telephone counselling. All NMAA counsellors have breastfed their own children, undergone a period of training followed by assessment, and are prepared to work voluntarily. The NMAA is supported by a panel of expert advisers from state health authorities and specialists in infant health and nutrition. They publish a variety of educational booklets on breastfeeding, a newsletter and, since 1982, have produced a twice-yearly professional publication entitled *Breast-feeding Review* [91].

The NMAA has opened the Lactation Resource Centre to collect, file and disseminate the latest information on breastfeeding. In addition to being an educational center, it also coordinates and encourages a wide range of research projects on breastfeeding [91].

The NMAA has been widely acknowledged for its positive contribution to breastfeeding in Australia. In the report where the original turnaround in Victorian breastfeeding trends was noted, acknowledgment was given to the NMAA for its contribution. Smibert [13, 15] echoes this acknowledgment and states that the increasing acceptance of breastfeeding was stimulated by mothers themselves, not the health professionals, and was perpetuated by the NMAA. Hartmann et al. [95–97] also agree that the increases in breastfeeding rates seen in Western Australia were largely due to the work of the NMAA.

In 1974, the first two books on breastfeeding in Australia were published. Both were written by mothers who were encouraging other women to breastfeed. One author [98] was a member of NMAA and the other a member of Parent Centres Australia [99].

In 1984, the International Lactation Consultant Association was formed. An affiliation was formed in Australia and it was incorporated in 1987. Support for the WHO Code is written into its constitution. The role of the ALCA complements that of the lay breastfeeding counsellors of the NMAA. Lactation consultants work in hospitals, clinics, and the community emphasizing the benefits of breastfeeding and the prevention of breastfeeding problems. Most certified lactation consultants are midwives or maternal and child health nurses with specialist skills. All consultants must reach an international competency level. In July 1991, there were 293 certified lactation consultants in Australia; 18% of the world total. All consultants must be recertified every 5 years to ensure continued professional growth and maintenance of standard competencies. The ALCA actively supports the Lactation Resource Centre and liaises with NMAA wherever possible [91, 100].

Summary of Support for Breastfeeding

Formal national government support for breastfeeding began with an NHMRC statement on feeding of infants and young children in 1976. 'Increase breast-feeding' was adopted as one of Australia's dietary goals in 1979. Although Australia voted in favor of the WHO Code in 1981, it remains one of many countries which has not fully adopted the Code. The NHMRC released a statement in support of the Code in 1981 and guidelines for its implementation in 1985. Responsibility for implementation of the Code lies mainly with the states and territories. A breastfeeding target was included among the national health targets in 1986 and 1988. Following the World Summit for Children, Australia formally ratified the 'Plan of Action' in May 1991. Discussions on the Baby Friendly Hospital Initiative implementation in Australia have commenced. Superseding the voluntary industry code, Australian Agreement, an agreement on 'Marketing in Australia of Infant Formulas: Manufacturers and Importers' was signed in May 1992. An independent panel is to be set up to monitor implementation of the agreement.

To date, the most glaring deficiency in national government support is the lack of adequate routine mechanisms for monitoring rates of breastfeeding and implementation of the WHO Code. The absence of operational monitoring systems makes it impossible to assess attainment of specified breastfeeding prevalence goals and compliance with the WHO Code. The present division between federal and state/territory responsibilities adds to the difficulties encountered. However, monitoring is an area in which the national government should have a clear role but which, as yet, it has not picked up.

Activities of nongovernment organizations in support of breastfeeding have undoubtedly been the most important avenues of support for breastfeeding in Australia. The NMAA was founded in 1964 and has played a most important role since then in promoting breastfeeding. In addition to its counselling role with individual mothers, NMAA and other nongovernment organizations have lobbied for full implementation and monitoring of the WHO Code, kept records of Code violations, raised issues about maternity hospital practices, and conducted public education.

Conclusions

In the last 40 years, the level and pattern of breastfeeding in Australia has changed markedly. From the information available, it seems that in 1950, approximately 50% of infants were fully breastfed at 3 months of age. By the early 1970s this figure had dropped to approximately 20%. While the proportion of fully breastfed infants at 6 months of age was always approximately 10% lower, the trend was similar, with the lowest point being reached in the early 1970s at approximately 10%. From this low point, prevalence rates began to rise rapidly. By the early 1980s, they had returned to their 1950 levels. Rates continued to climb until 1985, at which point they appear to have levelled.

The avant-garde in these changes, both the decrease and the subsequent increase in breastfeeding rates, has been better-educated and higher-income women.

The most important influences on the increased levels of breastfeeding in the last 20 years have been women themselves, attitude changes in society, and nongovernment organizations such as the NMAA. Through the 1970s, there was little interest in the promotion of breastfeeding from either state or federal governments. It was a decade after the turnaround in prevalence rates that the WHO Code was adopted and federal government (through the NHMRC) and state governments began to actively support the return to breastfeeding. In effect, government activity has confirmed and reinforced what was already occurring.

This raises the question as to why it took 10 years for governments to respond. Australian politicians have taken little interest in public health until very recently. Their interest in health has been to contain costs in the hospital sector – a relatively straightforward issue to understand – rather than tackle the conceptually difficult issues of public health. The inherently conservative public service, frequently short on technical expertise, has not been recognized as an innovator or promoter of new trends. In relation to breastfeeding, these factors have meant that Australian politicians and public servants have been more influenced by international trends than by advice from within Australia.

Nevertheless, there is an important role for government in the current situation. Perhaps the most important is to establish and maintain a good mechanism for monitoring the rates of breastfeeding in Australia with a capacity to distinguish between ethnic and socioeconomic groups. It is only if this is established with standardized definitions and collection procedures that we will be able to know what is happening across the country and over time. This is essential, if we are to be able to measure progress towards national targets. At the moment, this monitoring capability should be given higher priority than the introduction of new programs.

A second role for government is ensuring full adoption of international initiatives already embraced. Continued development of agreements in line with the WHO Code are needed. Future agreements must delineate the responsibilities of the health care system, retailers and pharmacies, and bottle and teat manufacturers and importers. With their adoption, particular emphasis needs to be given to improving the monitoring of both implementation and compliance. Nationwide incorporation of the Baby Friendly Hospital Initiative into maternity facility accreditation procedures deserves support.

Thirdly, the most important issue in the promotion of breastfeeding in Australia is to increase the duration of breastfeeding once it has been initiated. At the moment our understanding of the reasons why women cease or never commence breastfeeding is very inadequate and, as a result, the design of programs to intervene is hampered. Identification of critical points, times when many women give up breastfeeding, and why they do so is crucial in order to improve breastfeeding promotion and support programs. Some of these critical points would seem to be the immediate posthospital period when a mother returns home with no social or medical support, times when the mother perceives that her milk supply is inadequate, and the return to a work place with little support for breastfeeding [53]. It is also necessary to better understand problems faced by low-income, Aboriginal and immigrant women, such as difficulty in accessing services and information, institutional intimidation, lack of confidence, and isolation. Government and nongovernment funding needs to be allocated towards understanding the barriers which prevent prolonged breastfeeding. When these are well-understood for all subgroups, then appropriate measures can be well-targeted.

References

1   Hitchcock NE: Part 3: Australian Aborigines and recent migrants. Aust J Nutr Diet 1989;46: 108–111.
2   Harrison L: Food, nutrition and growth in Aboriginal communities; in Reid J, Trompf P (eds): The Health of Aboriginal Australia. Sydney, Harcourt Brace Jovanovich, 1991, pp 151–172.
3   Gracey M: Maternal health, breast-feeding and infant nutrition in Australian Aborigines. Acta Paediatr Jpn 1989;31:377–380.
4   Hitchcock NE: Infant feeding in Australia: An historical perspective. Part 1: 1788–1900. Aust J Nutr Diet 1989;46:62–66.
5   Thearle MJ: Infant feeding in colonial Australia 1788–1900. Aust Paediatr J 1985;21:75–79.
6   Gandevia B: Tears Often Shed: Child Health and Welfare in Australia from 1788. Pergamon, 1978.

7   Coy JF: Hazards in infant feeding. Food Technol Aust 1971;23:382–387.
8   Hitchcock NE: Infant feeding in Australia: An historical perspective. Part 2: 1900–1988. Aust J Nutr Diet 1989;46:102–108.
9   Davis A: Infant mortality and child saving: The campaign of women's organisations in Western Australia, 1900–1922; in Hetherington P (ed): Childhood and Society in Western Australia. Perth, University of Western Australia Press, 1988, pp 161–173.
10 Reiger K: Women's labour redefined: Child-bearing and rearing advice in Australia, 1880–1930s; in Bevege M, James M, Shute C (eds): Worth her Salt: Women at Work in Australia. Sydney, Hale & Iremonger, 1982, pp 72–82.
11 Coy JF, Lewis IC, Mair CH: Tasmanian Infant Feeding Survey. Med J Aust 1970;00:132–134.
12 Coy JF, Longmore EA, Mair CH, et al.: Tasmanian Infant Feeding Survey, 1974. Aust Paediatr J 1976;12:171–175.
13 Smibert J: A history of breastfeeding: With particular reference to the influence of NMAA in Victoria. Breastfeed Rev 1988;1(12):14–19.
14 Manderson L: Infant feeding practice in Australia: A review of research of recent trends. Aust J Early Child 1989;14:30–35.
15 Smibert J: Two books on breast feeding published in Australia. Med J Aust 1975;2:954–955.
16 Mobbs GA, Mobbs EJ: Breast feeding – A suggestion. Med J Aust 1972;2:392.
17 Manderson L: To nurse and to nurture: Breastfeeding in Australian society; in Hull V, Simpson M (eds): Breastfeeding, Child Health and Child Spacing: Cross-Cultural Perspectives. London, Croom Helm, 1985, pp 162–187.
18 Mobbs EJ, Mobbs GA: Breast feeding – Success (or failure) due to attendants and not prevailing fashion. Med J Aust 1972;1:770–772.
19 Newton DB: The future of breast feeding. Med J Aust 1966;2:842–844.
20 Minchin M: Infant formula: A mass, uncontrolled trial in perinatal care. Birth 1987;14:25–34.
21 Simbert J: The return to breast feeding. Med J Aust 1978;2:533.
22 Nursing Mothers' Association of Australia: Compendium of Australian Breastfeeding Statistics. Nunawading, Nursing Mothers' Association of Australia, 1991.
23 Scott E: The statistical decline of breastfeeding in the first three months. Breastfeed Rev 1988;1:65.
24 Child, Adolescent and Family Health Services, South Australia: Infant Feeding Patterns: An Analysis of CAFHS Feeding Survey. No 87/008R. Adelaide, Child, Adolescent and Family Health Services, 1987.
25 Palmer N: Breast-feeding – The Australian situation. J Food Nutr 1985;42:13–18.
26 Australian Bureau of Statistics: 1989–90 National Health Survey Users' Guide. Catalogue No 4363.0. Canberra, Commonwealth Government Printer, 1991.
27 Australian Bureau of Statistics: 1989–90 National Health Survey Summary of Results: Australia. Catalogue No 4364.0. Canberra, Commonwealth Government Printer, 1991.
28 Coy JF, Lowry RK: Tasmanian Infant Feeding Survey 1979–1980. Hobart, Department of Health Services, 1980.
29 Hitchcock NE, Coy JF: Infant-feeding practices in Western Australia and Tasmania: A joint survey, 1984–1985. Med J Aust 1988;148:114–117.
30 Hitchcock NE, Coy JF: The growth of healthy Australian infants in relation to infant feeding and social group. Med J Aust 1989;150:306–311.
31 Hitchcock NE, McGuiness D, Gracey M: Growth and feeding practices of Western Australian infants. Med J Aust 1982;1:372–376.
32 Lawson JS: The return to breastfeeding. Med J Aust 1978;2:229–230.
33 Allen J, Heywood P: Infant feeding practices in Sydney 1976–1977: Interim report. Proc Nutr Soc Aust 1977;2:84.
34 Allen J, Heywood P: Infant feeding practices in Sydney 1976–77. Aust Paediatr J 1979;15:113–117.
35 Borda EC, Feeney EM, Morris MM, et al.: Current patterns of breast feeding in a New South Wales maternity hospital. Med J Aust 1978;2:250–253.
36 Webb K: Survey of Infant Feeding Practices in Western Metropolitan Health Region hospitals. Sydney, Metropolitan Health Region Department of Health, 1985.

37 Cox J: The antenatal and perinatal characteristics of socio-economically depressed Caucasians. Aust NZ J Obstet Gynaecol 1981;21:20–23.

38 Eaton-Evans J, Townsend B, Dugdale A: The milk feeding of infants in south-east Queensland 1972–1982. J Food Nutr 1985;42:137–142.

39 Eaton-Evans J, Dugdale AE: Effects of feeding and social factors on diarrhoea and vomiting in infants. Arch Dis Child 1987;62:445–448.

40 Tudehope D, Bayley G, Munro D, et al.: Breast feeding practices and severe hyperbilirubinaemia. J Paediatr Child Health 1991;27:240–244.

41 Hankin ME: Infant feeding. Food Nutr Notes Rev 1965;22:47–55.

42 Boulton TJC, Coote LM: Nutritional studies during early childhood. I: Energy and nutrient intake. Aust Paediatr J 1979;15:72–80.

43 Boulton TJC, Coote LM: Nutritional studies during early childhood. II: Feeding practices during infancy, and their relationship to socio-economic and cultural factors. Aust Paediatr J 1979;15: 81–86.

44 Newton DB: Breast feeding in Victoria. Med J Aust 1966;2:801–804.

45 Williams HE, Carmichael A: Nutrition in the first year of life in a multi-ethnic poor socio-economic municipality in Melbourne. Aust Paediatr J 1983;19:73–77.

46 Plovanic P, Lumley J: Breastfeeding patterns in inner city Melbourne. Nurs Mothers Assoc Aust Newslett 1984;20(9):12–15.

47 Owles EN, Hitchcock NE, Gracey M: Feeding patterns of Australian infants: Birth to one year. Hum Nutr 1982;36A:202–207.

48 Hitchcock NE, Owles EN, Gracey M, et al.: Development of eating and drinking patterns in the first two years of life. J Food Nutr 1983;40:161–164.

49 Hitchcock NE, Gracey M, Gilmour AI, et al.: Nutrition and Growth in Infancy and Early Childhood: A Longitudinal Study from Birth to Five Years. Monogr Paediatr Basel, Karger, vol 19, 1986.

50 Reynolds B, Hitchcock NE, Coveney J: A longitudinal study of Vietnamese children born in Australia: Infant feeding, growth in infancy and after five years. Nutr Res 1988;8:593–603.

51 Auerbach KG, Renfrew MJ, Minchin M: Infant feeding comparisons: A hazard to infant health? J Hum Lact 1991;7:63–71.

52 Labbok M, Krasovec K: Toward consistency in breastfeeding definitions. Breastfeed Rev 1991;2:121–124.

53 Winikoff B: Issues in the design of breastfeeding research. Stud Fam Plann 1981;12:177–184.

54 Mathers CD: Australia's Health Goals and Targets: Data Requirements and Recommendations for Review: Report to the National Better Health Program Management Committee. Canberra, Australian Institute of Health, 1990.

55 Eaton-Evans J, Dugdale AE: Recall by mothers of the birth weights and feeding of their children. Hum Nutr 1986;40A:171–175.

56 Bundrock V: Priorities for national health statistics. Breastfeed Rev 1990;2:60–65.

57 Siskind V, Schofield F, Rice D, et al.: Breast cancer and breastfeeding: Results from an Australian case-control study. Am J Epidemiol 1989;130:229–236.

58 Boulton TJC, Flavel SE: The relationship of perinatal factors to breastfeeding. Aust Paediatr J 1978;14:169–173.

59 Lilburne AM, Oates RK, Thompson S, et al.: Infant feeding in Sydney: A survey of mothers who bottle feed. Aust Paediatr J 1988;24:49–54.

60 Redman S, Booth P, Smyth H, et al.: Preventive health behaviours among parents of infants aged four months. Aust J Public Health 1992;16:175–181.

61 Baghurst KI: Infant feeding – Public health perspectives. Med J Aust 1988;148:112–113.

62 Coyne T, Dowling M: Infant feeding practices among Aboriginals in rural New South Wales. Proc Nutr Soc Aust 1978;3:91.

63 Dugdale AE: Infant feeding, growth and mortality: A 20-year study of an Australian Aboriginal community. Med J Aust 1980;2:380–385.

64 Phillips FE, Dibley MJ: A longitudinal study of feeding patterns of Aboriginal infants living in Perth, 1980–1982. Proc Nutr Soc Aust 1983;8:130–132.

65 Gracey M, Murray H, Hitchcock NE, et al.: The nutrition of Australian Aboriginal infants and young children. Nutr Res 1983;3:133–147.

66 Matthews M, Manderson L: Infant feeding practices and lactation diets amongst Vietnamese immigrants. Aust Paediatr J 1980;16:263–266.

67 Manderson L, Matthews M: Vietnamese attitudes towards maternal and infant health. Med J Aust 1981;1:69–72.

68 Ward BG, Pridmore BR, Cox LW: Vietnamese refugees in Adelaide: An obstetric analysis. Med J Aust 1981;1:72–75.

69 Breakey J: Possible causes of dietary changes in Vietnamese migrants in Australia. Proc Nutr Soc Aust 1983;8:56–63.

70 Bailey VF, Sherriff J: Reasons for the early cessation of breast-feeding in women from lower socio-economic groups in Perth, Western Australia. Aust J Nutr Diet 1992;49:40–43.

71 McIntyre E: Early cessation of breast-feeding. Aust J Nutr Diet 1992;49:38–39.

72 Leeson R: Weight gains in the breastfed baby. Breastfeed Rev 1992;2(5):246–250.

73 Phillips V: Supporting breastfeeding beyond hospital discharge. Breastfeeding: Perinatal and Postpartum Issues Conf. Brisbane, 25–27 Aug. 1991.

74 Minchin MK: Infant feeding. Med J Aust 1988;148:604.

75 National Health and Medical Research Council: Dietary Guidelines for Australians. Canberra, Australian Government Publishing Service, 1992.

76 Australian Bureau of Statistics: Census of Population and Housing, 30 June 1986. Census 86 – Australia in Profile: A Summary of Major Findings. Catalogue No 2502.0. Canberra, Commonwealth Government Printer, no date.

77 Glezer H: Maternity Leave in Australia: Employee and Employer Experiences: Report of a Survey. Aust Inst Fam Stud Monogr Melbourne, Australian Institute of Family Studies, No 7, 1988.

78 WHO: Infant and Young Child Feeding: Current Issues. Geneva, World Health Organization, 1981.

79 WHO: International Code of Marketing of Breast-Milk Substitutes. Geneva, World Health Organization, 1981.

80 WHO/UNICEF: Protecting, Promoting and Supporting Breast-feeding: The Special Role of Maternity Services. Geneva, World Health Organization, 1989.

81 United Nations Administrative Committee on Coordination – Subcommittee on Nutrition: The Innocenti Declaration in the protection, promotion and support of breastfeeding. Extracts from SCN News 1991;7:4–5.

82 UNICEF: Take the Baby-Friendly Initiative: A Global Effort with Hospitals, Health Services and Parents to Breastfeed Babies for the Best Start in Life. New York, UNICEF, no date.

83 Anonymous: An initiative that responds to landmark declarations. Baby-Friendly Hosp Init Newslett 1992;Jun:1.

84 Department of Community Services and Health, National Health and Medical Research Council: Nutrition Policy Statements: 1990 Edition. Canberra, Department of Community Services and Health, 1990.

85 Better Health Commission: Looking Forward to Better Health. Canberra, Australian Government Publishing Service, Vol 2, 1986.

86 Health Targets and Implementation (Health for All) Committee: Health for All Australians: Report to Australian Health Ministers. Canberra, Australian Government Publishing Service, 1988.

87 Australian Institute of Health: National Better Health Program Improved Nutrition: Monitoring Targets towards 2000. A Statistical Report to the Project Planning Team. Canberra, Australian Institute of Health, 1989.

88 National Health and Medical Research Council: Report of the Working Party on Implementation of the WHO International Code of Marketing of Breast-Milk Substitutes. Canberra, Australian Government Publishing Service, 1985.

89 Anonymous: WHO Code in Australia. Breastfeed Rev 1992;2(5):241.

90 Dietitians' Association of Australia: The marketing in Australia of infant formulas: Manufacturers and importers. Diet Assoc Aust Newslett 1992;64:34–35.

91 Brodribb W: Breastfeeding Management in Australia: A Reference Study Guide. Nunawading, Nursing Mothers' Association of Australia, 1991.

92 Mackerras D: Implementation of the WHO International Code of Marketing of Breast-Milk Substitutes in Australia. Proc 11th Natl Conf Diet Assoc Aust 1992, pp 38.

93 Shamis S, Thompson S, Cooper R, et al.: Draft policy statements: Breastfeeding. In Touch 1992;9(3):8–9.

94 Anonymous: An advocate in Australia. Baby-Friendly Hosp Init Newslett 1992;Aug:3.

95 Hartmann PE, Rattigan S, Prosser CG: Studies on breastfeeding and reproduction in women. J Food Nutr 1982;39:46–50.

96 Hartmann PE, Kulski JK, Rattigan S, et al.: Lactation in Australian women. Proc Nutr Soc Aust 1980;5:104–110.

97 Hartmann PE: Lactation and reproduction in Western Australian women. J Reprod Med 1987;32:543–547.

98 Phillips V: Successful Breast Feeding. Hawthorn, Nursing Mothers' Association of Australia, 1974.

99 O'Brien P: Discovering Childbirth and the Joy of Breastfeeding. Sydney, Antipodean, 1974.

100 Phillips V: Lactation consultants and voluntary breastfeeding counsellors: Complementary roles or conflict? Breastfeed Rev 1990;2(2):92–94.

Margaret Lund-Adams, Nutrition Program, University of Queensland,
Level 3 Edith Cavell Building, Royal Brisbane Hospital, 4029 Brisbane,
Queensland (Australia)

Simopoulos AP, Dutra de Oliveira JE, Desai ID (eds): Behavioral and Metabolic
Aspects of Breastfeeding. World Rev Nutr Diet. Basel, Karger, 1995, vol 78, pp 114–127

....

Breastfeeding in Korea

*Sook He Kim, Woo Kyoung Kim, Kyoung Ae Lee, Yo Sook Song,
Se Young Oh*

Department of Foods and Nutrition, Ewha Woman's University, Seoul, Korea

Contents

Infant Feeding Practices in Korea

Until the 1950s, breastfeeding was the sole means of providing nutrients to infants up to 6 months of age in Korea unless the mothers suffered from either health problems or insufficient breastmilk. The most common substitute for breast milk was rice gruel or rice soup. It was quite common for children 3–5 years of age to be partially breastfed. At that time, bottle feeding was a foreign concept and substitute milk was rarely available. Generally, bottle-fed children were considered to be nutritionally at risk compared to those who were breastfed.

At the beginning of the 1960s, bottle feeding was practiced only among women from the upper-middle class, especially by those with professions. They depended on imported foreign brands of powdered milk such as Morinaga from Japan and Villak from Australia – even though their prices were 10-fold higher than those for domestic products – mainly because the Korean food-processing industry was poorly developed at that time. Even though there were 11 different Korean brands of powdered milk available, none were considered appropriate.

Since the 1960s, Korean society has been greatly influenced by Western culture due to rapid industrialization. Educational and employment opportunities for women have increased, taking them out of the home. With the rapid development of food-processing industries since the 1960s, the general population had easier access to powdered milk. In the late 1960s and early 1970s, as good-quality domestic powdered milks became widely available, bottle feeding was no longer exclusively practiced by the upper-middle class and became a common feeding practice, particularly in urban areas. The broader social participation of women and increased availability of powdered milks appeared to promote bottle feeding in urban areas. Yet, in rural areas, most children were still breastfed.

In the 1950s and 1960s, a 'baby boom' led to a rapid population growth of 2% annually. This prompted the Korean government to promote family planning, resulting in a decrease in the rate down to 1% in the 1970s [1]. Family planning and rapid industrialization contributed to an increase in nuclear families in Korean society. In nuclear families, traditional child-rearing practices involving breast-feeding were less preserved, and bottle feeding appeared to be an attractive practice.

Before the mid-1960s, with the exception of abnormal cases, mothers delivered their babies at home with the help of midwives or their families not only in rural but also in urban areas. Hospital delivery, which had rarely been practiced before except for those from the upper classes became established in Korea in the mid 1960s. In hospitals, infants were bottle fed immediately after birth, and in many cases they continued to be bottle fed after discharge.

Factors Affecting Infant Feeding Practices

According to the survey conducted by Ewha Woman's University for 930 urban mothers in 1984 [2], an insufficient flow of breast milk was the main factor inhibiting breastfeeding, followed by employment (table 1) [2]. Those mothers that preferred breastfeeding did so because of its convenience, for nutritional and emotional considerations, or as a consequence of family tradition and peer group and husband's influences (table 2) [3]. From a consideration of the statements made by Korean mothers, it appears that breast milk was considered to be nutritionally superior to infant formula and bottle feeding was practiced mainly because of insufficient breast milk.

Recently, breastfeeding has been publicly promoted in Korea, particularly for immunological and psychological reasons. As a result, more urban infants had colostrum in 1991 than in 1979 (table 3) [4].

Table 1. Factors influencing infant formula or mixed-feeding practices in urban areas [2]

Factor	Number (%)
Not enough breast milk	384 (41.3)
Working mother	147 (15.8)
Baby refused breast milk	62 (6.7)
Breast nipple problems	61 (6.6)
No particular preference	59 (6.3)
Mother's health problems	52 (5.6)
Mother refused breastfeeding	51 (5.5)
Others	114 (12.2)

Table 2. Breastfeeding determinants, correlation coefficients [3]

Variable	Correlation coefficient (r)	
Breast milk is nutritionally better	0.2071	$p < 0.001$
Can be fed only breast milk until 6 months without any other supplementation	0.1980	$p < 0.001$
Is convenient to feed baby without any preparation	0.4620	$p < 0.001$
During pregnancy mother decided to breastfeed	0.3031	$p < 0.001$
Immunological considerations	0.1453	$p < 0.01$
Previous experience of breastfeeding	0.1847	$p < 0.01$
Better for mental stability and emotional satisfaction	0.1470	$p < 0.01$
Family tradition	0.1096	$p < 0.05$
Peer group influence	0.1163	$p < 0.05$
Husband's influence	0.0836	$p < 0.05$

Table 3. Colostrum-suckling status of newborns (%) in urban and rural areas [4]

Year	Urban	Rural
1979	72.4	
1983	72.8	94.7
1991	81.8	94.1

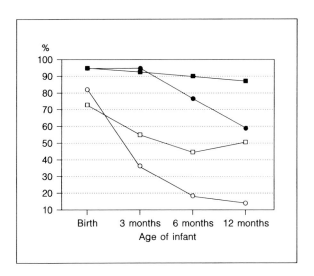

Fig. 1. Cumulative continuation rates for breastfeeding. ■ = Rural, 1983; ● = rural 1991; □ = urban, 1983; ○ = urban, 1991.

Table 4. Infant feeding practices in Korea (%) [4–10]

Type of feeding	Area	1967–1968	1974–1978	1981–1986	1987–1990
Breast	rural	95.6	82.0	73.5	58.8
	urban	60.4	51.7	28.5	22.7
Bottle	rural	4.4	8.0	10.3	5.9
	urban	16.5	19.8	41.0	63.6
Mixed	rural	0.0	10.0	12.3	35.3
	urban	23.1	28.5	30.5	13.6

Changes in Infant Feeding Practices

Korean infants are generally fed by the following methods: (1) breastfeeding; (2) formula feeding, and (3) mixed feeding with breast milk and formula. These feeding practices are influenced by many factors including residential area, infant's age, mother's education and employment.

As shown in table 4, fewer children were breastfed in the late 1980s than in the late 1960s [4–10]. It should be noted that, here, breastfeeding is defined as giving breast milk regardless of duration, frequency and amount. Infants were more frequently breastfed in rural areas than in urban areas because of easy access to good-quality formula milk in the latter, as mentioned before.

Table 5. Rates (%) of breastfeeding at various infant ages in relation to the mother's educational background [4, 11]

Infant's age months	Year	Mother's educational background			
		primary school	middle school	high school	college
3	1983	90.3	93.8	81.6	57.2
	1991	100.0	88.9	57.9	14.3
6	1983	89.6	91.3	73.0	44.3
	1991	100.0	77.7	26.4	14.3
12	1983	88.3	86.2	62.4	31.7
	1991	50.0	66.6	26.4	0.0

Table 6. Rates (%) of breastfeeding comparing working and nonworking mothers [11]

	At birth	Infant's age			
		3 months	6 months	9 months	12 months
Working mother	76.6	62.6	50.2	44.8	43.5
Nonworking mother	92.4	86.8	81.6	77.0	75.4

Breastfeeding rates as infants grew decreased between 1983 and 1991 in both urban and rural areas (fig. 1) [4, 11].

The mother's education is negatively associated with breastfeeding practices in Korea. This association is more prominent now than in the past (table 5) [4, 11]. Mothers with higher education are more likely to have professional jobs than those with less education, and thus less time for childcare. Working mothers breastfed their infants less than nonworking mothers (table 6) [11].

Weaning Foods

The introduction of weaning foods is related to a series of factors including feeding practices, region and infants age. It was reported that fruit juices were provided to infants less than 3 months old. According to data collected in 1991, about 20% of breastfed infants were introduced to weaning foods at 3 months, and 73% at 6 months (table 7) [4, 11].

Table 7. Changes over time in the introduction of weaning foods at different ages compared by feeding patterns [4, 11]

Feeding	Breastfeeding			Bottle feeding		
	3 months	6 months	12 months	3 months	6 months	12 months
1983	2.5	45.0	90.4	18.8	72.1	96.4
1991	20.0	73.3	100.0	46.7	100.0	100.0

Values are percentages.

Table 8. Rates (%) of introduction of weaning foods at different infant ages in rural and urban regions [4, 11]

Area	Year	3 months	6 months	9 months	12 months
Urban	1983	14.1	73.7	93.7	95.3
	1991	50.0	100.0		
Rural	1983	4.7	37.4	74.8	92.5
	1991	5.9	70.6		100.0

Bottle-fed infants were weaned earlier than breastfed infants, partly because Korean mothers think breastmilk is nutritionally better than formula milk [4, 11]. At 3 months, about 6% of rural infants and 50% of urban infants were weaned. At 6 months, these proportions increased: 100% of urban infants and 70.6% of rural infants received weaning foods, and all of the infants in the study were weaned at 12 months. In 1983, weaning foods were introduced to infants in both urban and rural areas at later ages than in 1991, and in both cases urban infants were weaned at earlier ages than their counterparts in the countryside (table 8) [4–11].

The regional differences observed in these studies are possibly related to economic status since urban people were economically better-off than rural people in Korea at these times. By the same token, earlier introduction of weaning foods in 1991 than in 1983 could be associated with the economic status of Korea which was better in 1991 than in 1983. The data also suggest that the ages at which infants are weaned is negatively related to economic status. Descriptive results on ages at which infants were weaned and mother's educational levels are presented in table 9 [4, 11].

Infants of mothers with college education were weaned 6 months earlier than of those with primary school education. It should be noted, however, that there

Table 9. Rates (%) of introduction of weaning foods at various infant ages in relation to the mother's educational background [4, 11]

Infant's age months	Year	Mother's educational background			
		primary school	middle school	high school	college
3	1983	1.0	2.3	2.5	9.6
	1991		11.0	36.8	57.1
6	1983	29.9	36.9	57.6	73.3
	1991	75.0	77.7	89.4	100.0
12	1983	92.0	92.6	96.9	–
	1991	100.0	99.9	99.9	100.0

Table 10. Regional and age differences in intakes of weaning foods [4]

Weaning foods	Infant's age					
	0–3 months		4–6 months		7–12 months	
	rural	urban	rural	urban	rural	urban
Rice gruel	–	–	33.3	59.1	58.8	95.5
Cooked rice	–	–	91.7	77.3	100.0	100.0
Mashed potatoes	–	–	16.7	31.8	58.8	95.5
Fruit and vegetable juices	100.0	100.0	58.3	68.2	88.2	100.0
Chopped eggs, meat and fish	–	–	16.7	45.5	70.0	95.5
Milk and dairy products	–	–	16.7	23.5	23.5	13.6
Yoghurt	–	45.5	8.3	63.6	64.7	100.0
Commercial weaning foods	–	9.0	66.7	90.0	70.6	100.0
Breads, cookies	–	–	33.4	50.0	82.4	90.0
Others	–	–	0.0	0.0	5.9	13.6

Values are percentages.

was no difference in weaning period between mothers with college education and those with middle and high school education.

Depending on their ages, Korean infants received various kinds of supplementary foods. At 2–3 months, all of the weaning infants received fruit and vegetable juices. The proportion of carbohydrate foods (i.e. rice gruel and cooked rice) increased remarkably at 4–6 months, and protein foods such as eggs, fish and meat were added much more after 7 months (table 10) [4].

Table 11. Changes of intake rates of commercial weaning foods [4, 10, 12]

Year	Intake rates, %
1977–1979	23
1985	69.0
1991	66.7–100.0

Traditionally, it is believed in Korea that children should be introduced to solid foods when they have their first tooth. It has been thought that 6 months is an appropriate time for weaning. However, our data indicate that solid foods were provided to infants who were younger than 6 months. This may be related to Westernization and easy access to scientific information.

Recently, more infants tend to be given commercial weaning foods. Only 23% of infants had commercial weaning foods in the late 1970s, while 67–100% were receiving them in 1991 [4]. Before the age of 3 months, about 10% of infants received commercial weaning foods. After 4 months, however, 60% of infants had these foods in both rural and urban areas (table 11) [4, 10, 12].

Physical Growth, Motor and Mental Development

Adequate amounts of nutrients are necessary for infant growth and development. A study conducted by Song [4] showed that physical growth assessed by anthropometry during infancy was unrelated to feeding patterns except for changes in head circumference at 0–3 months.

During the first year of life, infant morbidity was examined using hospital records. Data collected in 1983 showed that breastfed infants tended to spend fewer days in hospital due to respiratory and gastrointestinal tract diseases (table 13) [11]. In Song's study, bottle fed rural infants spent more days in hospital than any other groups (table 14) [4].

When the type of feeding was divided into 3 groups as shown in table 15, there was no difference in the appearance of teeth regarding feeding type and residential area. On the other hand, when the infants were categorized into breastfeeding and others, breastfed urban infants had their first tooth at an earlier age than rural infants who were not breastfed (table 16) [4]. However, the developmental significance of these results is not clear.

Motor development was examined by head motor function, turning, sitting alone, standing with help and grasp. Mental development was assessed by own-name recognition. Urban infants who were breastfed passed the tests on

Table 12. Anthropometric values for infants 0–12 months of age

	Type of feeding	Age		
		0–3 months	4–6 months	7–12 months
Height increases, cm				
Rural	breast	9.6 ± 1.3^a	2.6 ± 1.2^a	8.4 ± 1.7^a
	bottle	10.0^a	2.7 ± 1.2^a	7.3 ± 2.3^a
	mixed	9.9 ± 1.9^a	2.6 ± 1.4^a	7.6 ± 1.1^a
Urban	breast	8.0 ± 2.3^a	2.0 ± 0.7^a	7.7 ± 1.8^a
	bottle	8.4 ± 2.7^a	2.7 ± 1.9^a	7.7 ± 2.1^a
	mixed	9.4 ± 1.4^a	2.8 ± 1.1^a	4.6^a
Weight gain, kg				
Rural	breast	3.8 ± 0.7^a	1.3 ± 2.4^a	1.5 ± 0.8^a
	bottle	3.4^a	1.4 ± 0.7^a	1.3 ± 1.0^a
	mixed	3.6 ± 0.4^a	1.9 ± 0.8^a	1.2 ± 0.4^a
Urban	breast	3.3 ± 0.8^a	1.6 ± 0.6^a	1.1 ± 0.8^a
	bottle	3.7 ± 0.8^a	1.5 ± 0.5^a	1.4 ± 0.7^a
	mixed	4.0 ± 0.6^a	1.9 ± 0.5^a	1.4^a
Increases in head circumference, cm				
Rural	breast	8.6 ± 0.9^b	7.0 ± 0.7^a	2.7 ± 0.8^a
	bottle	$9.0^{a,b}$	8.0 ± 0.8^a	2.7 ± 1.0^a
	mixed	8.9 ± 0.7^b	6.4 ± 0.8^a	2.9 ± 0.6^a
Urban	breast	6.3 ± 2.0^a	5.4 ± 0.9^a	2.7 ± 0.6^a
	bottle	6.4 ± 1.1^a	7.1 ± 0.5^a	2.9 ± 0.4^a
	mixed	$7.2 \pm 1.1^{a,b}$	6.2 ± 0.2^a	3.0^a

Values with different letters are significantly ($p < 0.05$) different from each other within ages and anthropometric measures.

head motor function, sitting alone and own name recognition at earlier ages than rural infants who were not breastfed (tables 17, 18). However, the former performed better only on the sitting alone test compared to breastfed rural infants. This suggests that breastfeeding practice could possibly reduce the difference in motor and mental development between urban and rural infants in Korea, although this study was unable to explain why. There were no differences in motor and mental developmental measures of infants in terms of feeding patterns within both rural and urban areas.

Since all the infants included in the study were not nutritionally at risk, feeding pattern is unlikely to affect physical growth. Differences in motor and

Table 13. Days spent in hospital due to illness of infants aged 0–12 months, surveyed in 1983 [11]

	Breast	Bottle	Mixed
0–2 months			
Respiratory	0.68	1.14	0.68
GI tract	0.31	0.54	0.52
3–5 months			
Respiratory	1.12	1.60	1.43
GI tract	0.65	0.93	0.72
6–8 months			
Respiratory	1.01	2.03	1.83
GI tract	0.91	0.56	1.86
9–10 months			
Respiratory	0.73	0.90	0.52
GI tract	0.55	0.77	1.52

GI = Gastrointestinal.

Table 14. Days spent in hospital due to cold or diarrhea, according to location, of infants aged 0–12 months, surveyed in 1991 [4]

	Breast		Bottle		Mixed	
	rural	urban	rural	urban	rural	urban
0–3 months						
Cold	0.4[a]	0.2[a]	8.0[b]	0.6[a]	0.3[a]	1.3[a]
Diarrhea	0.3[a]	0.0[a]	1.0[a]	0.2[a]	0.2[a]	0.0[a]
4–6 months						
Cold	2.6[a]	0.0[a]	4.3[a]	1.3[a]	1.8[a]	0.0[a]
Diarrhea	0.7[a]	0.5[a]	0.0[a]	0.6[a]	0.2[a]	0.0[a]
7–12 months						
Cold	1.8[a]	0.5[a]	1.3[a]	4.4[a]	5.0[a]	18.0[a]
Diarrhea	0.7[a]	0.0[a]	0.3[a]	0.7[a]	1.3[a]	0.0[a]

Values with different letters are significantly ($p < 0.05$) different from each other within rows.

Table 15. Appearance of infantile teeth (months): comparison of different feeding patterns of 0–3 months duration [4]

Area	Type of feeding	Teeth	
		lower front	upper front
Rural	breast	6.3 ± 1.9	7.6 ± 1.9
	bottle	8.0	9.0
	mixed	7.5 ± 0.7	8.8 ± 0.8
Urban	breast	5.8 ± 1.1	7.0 ± 0.9
	bottle	7.0 ± 0.2	8.2 ± 1.3
	mixed	7.0 ± 0.8	7.3 ± 0.6

Table 16. Appearance of infantile teeth (months): comparison of different feeding patterns of 0–6 months duration [4]

Area	Type of feeding	Teeth	
		lower front	upper front
Rural	breast	$6.1 \pm 1.9^{a,b}$	7.7 ± 1.3^{a}
	bottle and mixed	7.4 ± 0.6^{b}	8.8 ± 2.1^{a}
Urban	breast	5.0 ± 1.4^{a}	6.5 ± 0.7^{a}
	bottle and mixed	$6.9 \pm 1.0^{a,b}$	7.9 ± 1.2^{a}

Values with different letters are significantly ($p < 0.05$) different from each other within columns.

mental items between urban and rural infants suggest influences of environmental factors on these facets of development (tables 17, 18) [4].

Conclusion

Rapid industrialization, which also means rapid Westernization, has influenced women's role in Korean society. Since the 1960s, more women have been educated and employed outside the home. As a result, feeding patterns have gradually changed, and bottle feeding practices have become widespread among

Table 17. Ages (months) of achievement for motor and mental development of infants according to feeding patterns of 0–3 months duration [4]

Item	Type of feeding	Area	
		rural	urban
Head motor function	breast	$4.2 \pm 1.9^{a,b}$	3.0 ± 0.9^a
	bottle	$5.0^{a,b}$	$3.6 \pm 0.6^{a,b}$
	mixed	4.8 ± 0.7^b	$3.7 \pm 1.7^{a,b}$
Turning	breast	6.5 ± 1.9^a	5.4 ± 2.1^a
	bottle	5.0^a	5.9 ± 1.8^a
	mixed	5.5 ± 0.7^a	6.0 ± 0.8^a
Sitting alone	breast	8.4 ± 1.2^b	6.0 ± 1.4^a
	bottle	$10.0^{a,b}$	6.9 ± 1.5^a
	mixed	8.8 ± 2.3^b	$7.0 \pm 0.8^{a,b}$
Grasp	breast	7.7 ± 4.9^a	4.8 ± 2.2^a
	bottle	6.0^a	5.7 ± 2.3^a
	mixed	6.5 ± 0.7^a	7.0 ± 0.8^a
Standing with help	breast	10.8 ± 2.1^a	9.2 ± 0.8^a
	bottle	8.0^a	8.8 ± 1.7^a
	mixed	9.0 ± 1.4^a	10.0 ± 0.0^a
Own-name recognition	breast	$10.1 \pm 2.3^{a,b}$	7.5 ± 2.3^a
	bottle	$12.0^{a,b}$	$8.9 \pm 1.8^{a,b}$
	mixed	11.5 ± 3.5^b	$8.7 \pm 4.2^{a,b}$

Values with different letters are significantly ($p < 0.05$) different from each other within items and feeding types.

Korean mothers. Women with higher education or from economically better-off families, especially, have practiced bottle feeding more often than their counterparts in other classes.

Bottle feeding has probably become popular among mothers because they have not been able to spend enough time for breastfeeding, a consequence of practical considerations such as work schedules and insufficient breast milk. The studies indicate that most Korean mothers would prefer to breastfeed. They understand it to be good for their infant, both nutritionally and emotionally.

There is some suggestion that breastfeeding might be related negatively to the duration of morbidity, and positively to some of motor and mental items. Rather than feeding practices, economic factors are more likely to be related to the

Table 18. Ages (months) of achievement for motor and mental development of infants according to feeding patterns of 0–6 months duration [4]

Item	Type of feeding	Area	
		rural	urban
Head motor function	breast	$3.9 \pm 2.1^{a,b}$	2.5 ± 0.5^{a}
	bottle and mixed	5.1 ± 0.6^{b}	$3.5 \pm 0.9^{a,b}$
Turning	breast	6.5 ± 1.9^{a}	5.5 ± 0.5^{a}
	bottle and mixed	5.5 ± 0.6^{a}	5.8 ± 1.7^{a}
Sitting alone	breast	8.9 ± 1.5^{b}	5.5 ± 0.5^{a}
	bottle and mixed	8.4 ± 1.8^{b}	6.8 ± 2.0^{a}
Grasp	breast	8.2 ± 2.4^{a}	5.5 ± 0.5^{a}
	bottle and mixed	6.3 ± 0.6^{a}	5.9 ± 2.4^{a}
Standing with help	breast	11.2 ± 2.4^{a}	9.5 ± 0.5^{a}
	bottle and mixed	9.1 ± 1.0^{a}	9.0 ± 1.5^{a}
Own-name recognition	breast	$10.0 \pm 2.0^{a,b}$	6.5 ± 0.5^{a}
	bottle and mixed	11.4 ± 2.9^{b}	$8.8 \pm 2.1^{a,b}$

Values with different letters are significantly ($p < 0.05$) different from each other within items and feeding types.

growth and development of Korean infants whose nutritional status is generally not at risk.

It does not mean that breastfeeding is not a better option than bottle feeding for Korean infants and there is no doubt that breastfeeding should be promoted.

References

1   National Statistics Office: Trend of Population Projection. Seoul, National Statistics Office, 1991.
2   Kim CH: A Statistical Study of Feeding Trends; Master's thesis Ewha Woman's University, 1984.
3   Pope AM: The Influence of the Traditional Korean Childbearing Culture on Breast Feeding; PhD thesis Yonsei University, 1983.
4   Song YS: A Study on Nutritional Status of Pregnant Women and their Infant's Feeding Practice, Growth and Development until the First Year of Life: PhD thesis Ewha Woman's University, 1991.
5   Lee HK, Dok-ko YC, Whang WG: Studies on supplementary diets of infants and young children of the farm area in Korea. Kor J Nutr 1968;1:117–119.
6   Chung KB, Kwon HS: Distribution of different feeding patterns, and clinical observation of Korean infants. J Kor Pediatr 1975;18:55–67.

7   Ahn SJ: A study on the weaning status in middle cities in Korea. J Kor Home Econ 1977;15: 45–58.
8   Chun, SK: The improvements of infant nutritional status in Korea. J Kor Pediatr 1980;23:12–23.
9   Kweon EK, Tchai BS, Han JH: A study on breast feeding and socioeconomic factors in a part of Seoul and its rural area. J Kor Public Health 1985;11:17–28.
10 Pang HK, Kim KH, Park JO, Lee SJ: Present status and problems of weaning. J Kor Pediatr 1987;30:266–273.
11 Korean Institute for Population and Health: A longitudinal study of breast feeding practice during infancy: Patterns, correlates and health effects. Seoul, Korean Institute for Population and Health, 1983.
12 Chung YJ: Weaning practice for infants in Daejeon. Kor J Nutr 1979;12:23–30.

Sook He Kim, Department of Foods and Nutrition, Ewha Woman's University,
Seoul 120-750 (Korea)

Simopoulos AP, Dutra de Oliveira JE, Desai ID (eds): Behavioral and Metabolic
Aspects of Breastfeeding. World Rev Nutr Diet. Basel, Karger, 1995, vol 78, pp 128–138

....

Breastfeeding in China

Dong-Sheng Liu, Xibin Wang

Institute of Nutrition and Food Hygiene, Chinese Academy of Preventive Medicine,
Beijing, China

Contents

Introduction

Since time immemorial, breastfeeding has been a prerequisite for the survival of young infants. Human milk is the best food for full-term infants. It has some special characteristics which are matched to the nutritional needs and physiological limitations of infants. In addition, breast milk contains several factors which act to prevent infection and allergic diseases. Breastfeeding has an important contraceptive effect and stimulates a close interaction between mother and child. It also has economic advantages both for the family and at a national level.

In industrialized countries, the breastfeeding rate has declined over the last 50 years [1]. Only during the past decade has the rate begun to increase again in an encouraging way. In developing countries, the trend is towards a continuous decline, with no signs of improvement. Bottle feeding is introduced early by increasing numbers of women, particularly in urban areas but also among the rural population. The purpose of this paper is to present a brief overview of breastfeeding in China today.

Table 1. Infants feeding patterns in 20 provinces between 1983 and 1986 [2]

Feeding pattern	Number of cases			Percentage		
	urban	rural	total	urban	rural	total
Breastfeeding	27,434	25,097	52,521	48.8	75.1	58.6
Mixed feeding	20,352	7,729	28,081	36.2	23.1	31.3
Bottle feeding	8,403	584	8,997	15.0	1.8	10.1
	56,189	33,410	89,599	100	100	100

Table 2. The feeding patterns of 0- to 6-month-old infants in Chengdu [3]

Age (months)	Total	Breastfeeding		Mixed feeding		Bottle feeding	
		n	%	n	%	n	%
0	235	70	29.8	96	40.8	69	29.4
1	456	98	21.5	127	27.8	231	50.7
2	444	86	19.4	105	23.6	253	57.0
3	530	81	15.3	141	26.6	308	58.1
4	576	84	14.6	150	26.0	342	59.4
5	475	46	9.7	120	25.3	309	65.0
6	398	42	10.5	103	25.9	253	63.6
Total	3,114	507	16.3	842	27.0	1,765	56.7

Breastfeeding Epidemiology

China has begun to experience a rapid decline in the prevalence of breastfeeding in urban and periurban areas and there is concern that this is spreading to rural areas. Although quantitative data are not available, it is commonly recognized that breastfeeding was almost universal 30–40 years ago. According to a national study conducted between 1983 and 1986 [2] in 20 provinces with nearly 90,000 infants, the average breastfeeding rate among infants under 6 months of age was 48.8% in urban areas and 75.1% in rural areas (table 1). The mixed-feeding rate was 36.2% in urban and 23.1% in rural areas. In some cities such as Chengdu (Sichuan province), an investigation in 1985 [3] showed that the breastfeeding rate was only 16.3% and the bottle-feeding rate rose to 56.7% (table 2). While in rural

Table 3. Breastfeeding rate (%) in Beijing

	City	Suburbs	Average
1950s	81	75	88
1980s	22	61.5	41

areas, breastfeeding was the main feeding pattern, the bottle-feeding rate was only 1.78%. A survey of 2,287 infants in Beijing in 1982 revealed that only 22% of the infants in the city and 61.5% in the surrounding rural areas were breastfed by their mothers [4]. In contrast, retrospective study of infant feeding practices in Beijing showed that in the 1950s, the rate of breastfeeding in the city was about 81% and in the surrounding rural areas it was 95%. This means that in the past 30 years of so there has been a decline of approximately 40% in the incidence of breastfeeding in Beijing and its surrounding rural areas (table 3). As expected, the decline in the rate of breastfeeding was accompanied by a rise in the rate of mixed and bottle feeding, and in the requirement for milk substitutes for infants.

In urban areas, the negative factors currently influencing breastfeeding are: (1) a delay in the initiation of breastfeeding after delivery due to separation of baby and mother in the hospital; (2) mothers' early return to work and working places being too far from home; (3) the perception by many mothers that they have too little milk or that the milk is too dilute for their babies, and (4) easy accessibility to and the impetus to use either modified cow's milk or modern humanized infant formula as a substitute.

In rural areas, breastfeeding has been the traditional method for feeding infants. The factors contributing to this are: the shortage or absence of infant foods in many villages; the relative youth and good health of mothers, and the fact that working places are not far from the home. These factors account for the higher breastfeeding rate in rural than in urban areas.

Comparison of Infant Feeding Patterns

Opinions vary as to the effects of various feeding patterns on the growth of infants. The reports of Zhu et al. [5, 6] from Shanghai showed no significant difference in the effect of the three feeding patterns – breast, mixed and bottle – on the increment in weight and length of urban infants under 6 months of age (table 4). But the weight and length gain in the breastfed infants tended to be the lowest among urban infants of 6–12 months of age. This evidence suggests that after 6 months, breastfed infants grow at a slightly lower rate than bottle-fed infants. This may be due to the ease of overfeeding the baby with the bottle and

Table 4. The increment of weight and length of infants receiving various feeding patterns (mean ± SD) [4, 5]

	0–6 months			6–12 months		
	num-ber	weight (kg)	length (cm)	num-ber	weight (kg)	length (cm)
Urban infants						
Breast fed	30	4.78 ± 0.75	17.1 ± 1.48	29	1.65 ± 0.51	7.30 ± 1.30
Mixed fed	168	4.75 ± 0.75	17.3 ± 1.77	164	1.71 ± 0.41	7.82 ± 1.02
Bottle fed	49	4.86 ± 0.85	17.1 ± 1.98	46	1.73 ± 0.42	7.76 ± 1.26
Rural infants						
Breast fed	94	4.61 ± 0.77		88	1.31 ± 0.47	
Mixed fed	76	4.27 ± 0.84		65	1.46 ± 0.55	
Bottle fed	11	4.02 ± 0.33		11	1.90 ± 0.50	

Table 5. Feeding patterns and morbidity of infants during the first 4 months [7]

	Breastfeeding (n = 352)		Mixed feeding (n = 177)		Bottle feeding (n = 107)	
	n	%	n	%	n	%
Respiratory infection	78	22.2	55	31.1	51	47.7
Gastrointestinal tract disorders	42	11.9	38	21.5	33	30.8
Eczema	23	6.5	16	9.0	29	27.1
Bronchopneumonia	11	3.1	3	1.7	10	9.4
Other infections	17	4.8	3	7	14	13.1
Other noninfectious diseases	1	0.3	4	2.3	3	0.8

early introduction of supplementary foods. The rural babies who were exclusively breastfed from 0 to 6 months of age had a greater body weight gain than the mixed- and bottle-fed babies (table 4). However, from 6 to 12 months of age, the weight gain of the rural infants was lowest among the breastfed infants and highest in the bottle-fed infants. This suggests that the nutrients in breast milk might not be adequate to meet the physical-growth needs of infants in the second half of the first year. Therefore, to ensure normal physical growth after 6 months, it is important that proper weaning foods are provided, and that the education regarding scientific feeding is fully emphasized, especially in rural areas.

Table 6. Feeding pattern and incidence of rickets in infants during the first 4 months [8]

	Number of infants observed	First-degree rickets	
		n	%
Breast feeding	106	10	9.43
Mixed feeding	135	22	16.30
Bottle feeding	199	42	21.11

Table 7. Breast milk output of lactating mothers (g/day; mean $\pm$ SD) [12]

	1st month	3rd month	6th month
Urban mothers	689 $\pm$ 126	695 $\pm$ 210	547 $\pm$ 204
(n = 25)	(368 – 922)	(474 – 1,177)	(142 – 1,089)
Rural mothers	803 $\pm$ 158	803 $\pm$ 113	758 $\pm$ 209
(n = 25)	(550 – 1,224)	(607 – 1,105)	(337 – 1,151)

The values in parentheses are ranges. There are statistically significant differences between the mothers and between the two areas ($p < 0.01$).

The relationship between feeding patterns and infant morbidity was studied by Chen and Hu [7] in Wuhan (Hubei province). The survey showed that among 636 infants, 352 were breastfed (55.4%), 177 received mixed feeding (27.8%) and 107 were bottle fed (16.8%). The infants morbidity was 38.1, 52.0 and 68.2%, respectively. Table 5 shows that breastfed infants had lower incidence of respiratory infection, gastrointestinal tract infection, and eczema than mixed- and bottle-fed infants. In addition, Liu [8] has reported on the impact of various feeding patterns on the incidence of rickets among infants in Beijing (table 6). Apparently, the incidence was lower among the breastfed than the bottle-fed infants.

Breast Milk Volume and Composition

In China, many mothers are afraid that their breast milk is too dilute and too little for their babies. So the questions arise, how much milk can one expect from the average, healthy, well-nourished mother and what is the concentration of total energy and protein in the milk? There are many studies on the relationship between the nutritional status of lactating mothers and the output and nutrient composition of their milk. A cross-sectional study of 189 lactating women (80 in urban, 58 in suburban and 51 in rural areas) was reported by Wang et al. [9] from Beijing. The results showed that the average breast milk output of these mothers was similar to that of mothers in the developed countries, being 689, 784 and 778 g/day for the urban, suburban and rural mothers, respectively. Two other surveys [10, 11] showed that the milk secretions of rural lactating mothers were 845 and 951 g/day. A longitudinal study of the influence of dietary intakes on breast milk output and nutrient composition during the first 6 months after delivery was carried out in 1985 in urban and rural areas of Beijing by Liu et al. [12]. The results showed that the average daily breast milk output was in the range of 550–800 g during the first 6 months of lactation (table 7). The breast milk output remained quite constant during the first 3 months of lactation for both areas. But the figures for urban mothers were lower by about 150 g at 6 months, the difference being statistically significant ($p < 0.01$), while there was only a small decrease for the rural mothers. Generally the average figures are higher for rural than for urban mothers at any month of lactation. The amount of breast milk secreted by Chinese mothers is within the range for mothers in the Western industrialized countries. In comparison, mothers in other developing countries generally produce much less breast milk [13] (table 8).

In contrast with breast milk output, the energy and major nutrient content of breast milk is generally higher for urban than for rural mothers. The protein concentration for urban mothers is always higher at any month of lactation, although the differences are not significant [12] (table 9). The differences in the nutrient content of breast milk between the rural and urban mothers could be ascribed to differences in the quality of the diets of the two areas, but there is no statistically significant correlation. The Wang et al. [9] study also indicates that the nutrient status of these mothers has no direct influence on breast milk secretion and nutrient content. Generally speaking, it would appear that, except in severe maternal undernutrition, the concentrations of total energy and protein in breast milk are maintained at normal levels over a range of dietary intakes. Therefore, if the volume of breast milk is sufficient, we should encourage mothers to exclusively breastfeed their babies up to 6 months of age.

In addition, there are some studies on the essential inorganic elements and vitamins in human milk. Yin et al. [14] have surveyed the contents of vitamins A, B_1, B_2 and C in the breast milk of 152 lactating mothers, and calcium, phosphorus, magnesium, zinc, copper and iron in 132 lactating mothers in Beijing. Their results for vitamins were similar to those reported by Packard [15] and Guthrie [16] (table 10). The contents of the inorganic elements decreased as lactation advanced (table 11). Therefore, during the 3rd–4th months of breastfeeding, it is better to start some supplementary foods to add zinc, iron, calcium and vitamins A, D and B_1 so that infants can meet their daily requirement.

Table 8. Milk output (ml/day; mean) of women from different countries [9, 13]

Author	Country	Month of lactation	
		1	2
Hofvander	Sweden	660 (580–860)	755 (575–985)
Whitehead	UK	740 (480–1,059)	785 (380–1,235)
Chandra	Canada	–	–
Picciano	USA	606 (336–876)	601 (355–847)
Hennart	Zaire	517 (250–780)	–
Martinez	Mexico	–	577 (433–842)
Prentice	Gambia	–	677 (525–1,055)
Wang[a]	China (urban)	731 ± 123	711 ± 153
	(rural)	829 ± 259	763 ± 112

Values in parentheses are ranges.
[a]Mean ± SD.

Table 9. Nutrient contents (mean ± SD) of breast milk [12]

	1st month	3rd month	6th month
Urban Mothers			
Protein, g/100 g	1.3 ± 0.2	1.1 ± 0.1	1.0 ± 0.3
Fat, g/100 g	3.4 ± 1.4	3.3 ± 1.6	3.6 ± 1.8
Lactose, g/100 g	8.1 ± 0.9	8.1 ± 2.1	7.7 ± 0.5
Energy, kcal/100 g	69 ± 12	66 ± 15	67 ± 15
Rural Mothers			
Protein, g/100 g	1.2 ± 0.2	1.0 ± 0.1	1.0 ± 0.2
Fat, g/100 g	2.6 ± 1.2	2.6 ± 1.2	2.3 ± 1.0**
Lactose, g/100 g	7.9 ± 0.8	7.8 ± 0.6	8.1 ± 0.6
Energy, kcal/100 g	60 ± 11	59 ± 10	57 ± 10*

Statistically significant differences between the two areas at **$p < 0.01$, and *$p < 0.05$.

Jin et al. [17] investigated the lipid composition of breast milk in Beijing. Two hundred and twenty one milk samples were analyzed for their fatty acid composition. The main components were oleic acid (29–37%), palmitic acid (17–25%) and linoleic acid (12–25%), with 1–2% erucic acid. The breast milk of Chinese mothers has a higher linoleic acid content than than of breast milk in other countries. It might be influenced by the type of dietary fat eaten by the mothers. The essential fatty acids in breast milk seems to be adequate to the infant's needs. The cholesterol available from breast milk is about 70–85 mg/day. Zhao et al. [18]

3	4	5	6
780 (600–930)	795 (560–1045)	566 (170–950)	450 (50–1,145)
784 (280–1,114)	717 (210–1,091)	588 (183–1,020)	493 (135–906)
793 (651–935)	856 (658–1,054)	925 (701–1,149)	872 (602–1,124)
626 (392–860)	–	–	–
605 (390–920)	–	–	525 (180–1,080)
–	537 (455–663)	–	561 (432–850)
–	–	617 (355–885)	–
689 ± 172	668 ± 146	686 ± 168	–
754 ± 125	728 ± 240	771 ± 141	–

Table 10. The vitamin content in human milk

	A µg/100 g	B$_1$ µg/100 g	B$_2$ µg/100 g	Nicotinic acid µg/100 g	C mg/100 g
Yin et al. [14] (mean ± SD)	(n = 143) 10.9 ± 6.6	(n = 152) 13 ± 7	(n = 152) 48 ± 13	(n = 152) 185 ± 75	(n = 152) 4.7 ± 1.91
Packard [15] (range)	(n = 309) 4.5–67.8	(n = 279) 8–23	(n = 275) 19.8–79	(n = 271) 66–330	(n = 233) 0–11.2
Guthrie [16]	56.7	16	36	150	4.3

Table 11. Inorganic elements in breast milk during lactation [14]

Month	n	Calcium mg/100 g	Zinc µg/100 g	Copper µg/100 g	Iron µg/100 g
<1	9	30.7 ± 4.80	454.6 ± 70.4	45.2 ± 5.79	89.8 ± 21.9
1	20	31.7 ± 3.81	353.6 ± 109.9	40.7 ± 16.1	82.5 ± 21.9
2	24	30.1 ± 5.16	262.4 ± 126.8	25.3 ± 5.69	79.5 ± 23.5
3	24	30.0 ± 4.17	272.3 ± 81.1	24.1 ± 7.63	76.6 ± 30.1
4	22	29.3 ± 6.63	267.1 ± 98.8	21.5 ± 6.38	63.1 ± 16.4
5	23	27.8 ± 4.59	242.4 ± 76.8	20.2 ± 8.55	68.6 ± 9.65
6	10	25.9 ± 5.20	207.1 ± 120.5	21.6 ± 7.12	74.7 ± 24.9

Values are the mean ± SD.

Table 12. Amino acid content of breast milk in different areas of Beijing and in the USA [18]

	Beijing (mean ± SD)			USA (mean)
	urban (n = 26)	suburban	rural (n = 26)	
Aspartic acid	86.9 ± 13.4	93.0 ± 13.7	86.9 ± 13.8	82
Threonine	43.9 ± 7.5	46.7 ± 6.7	43.7 ± 6.6	46
Serine	44.6 ± 7.3	48.4 ± 7.7	43.5 ± 6.5	43
Glutamic acid	181 ± 21	194 ± 19	191 ± 23	168
Proline	93.5 ± 11.6	98.6 ± 9.8	98.6 ± 14.5	82
Glycine	22.0 ± 3.5	24.2 ± 4.9	22.0 ± 4.0	26
Alanine	37.6 ± 5.0	40.2 ± 6.4	38.0 ± 6.5	36
Cystine	20.4 ± 4.0	22.1 ± 3.8	20.6 ± 2.8	19
Valine	56.1 ± 7.2	58.5 ± 6.9	57.5 ± 8.7	63
Methionine	15.8 ± 2.8	16.5 ± 2.5	18.3 ± 2.4	21
Isoleucine	51.4 ± 7.0	52.7 ± 6.1	51.8 ± 6.9	56
Leucine	111 ± 18	114 ± 16	111 ± 18	95
Tyrosine	37.9 ± 4.6	40.0 ± 5.3	40.4 ± 4.9	53
Phenylalanine	35.4 ± 5.2	36.8 ± 6.4	35.4 ± 4.9	46
Histidine	27.2 ± 4.5	29.6 ± 4.9	28.8 ± 5.5	23
Lysine	67.6 ± 11.0	73.0 ± 9.5	68.2 ± 10.4	68
Arginine	36.6 ± 5.8	39.2 ± 8.1	37.0 ± 7.2	43
Tryptophan	17.0 ± 3.2	17.2 ± 2.5	16.4 ± 2.1	17

studied the protein and amino acid content of breast milk in Beijing (table 12). The results showed that the total amount of 18 amino acids in Beijing mothers' milk was similar to that in the milk of US mothers.

Weaning Practice

The patterns of feeding practice and the types of foods supplemented to 10- to 18-month-old infants are probably the most important determinants of growth in Chinese children. Very frequently, only rice or wheat porridge is used. Sugar is often added but rarely fat or oil. Most pediatricians feel that supplementary and home weaning foods in China do not contain enough calories. Data on weaning from seven provinces indicated that complete cessation of breastfeeding in rural areas takes place between 12 and 24 months of age. Forty to sixty percent of weanings were found to be abrupt, depending on the region of the country. A national study in 1985 [19] showed that about 65% of infants were receiving some cereals, usually in the form of porridge, by 5–6 months of age, about 35% were receiving fruits or vegetables by the same age, and about 35% had received some eggs by the age of 4 months. Less than 20% had some meat and only 5% had some bean products by 10 months of age. The Institute of Nutrition has estimated that for children under 1 year, the average daily intake of cereals was about 14–27 g/day and for sugar, 6–15 g/day. Altogether, these represent 150 kcal or only 18% of the infant's caloric need. It is, therefore, quite apparent that weaning and infant feeding practices in China need modification, including the development of home-based weaning foods that can be readily used by families for infants and young children.

Conclusion

From the reports reviewed, we can conclude that Chinese mothers, both urban and rural, have enough good-quality breast milk for their infants. They should be encouraged to trust in their ability to breastfeed. The promotion of breastfeeding is urgently needed in China and the Chinese government has been paying great attention to this problem. Recently, a symposium held by the Ministry of Public Health developed a project to promote breastfeeding. This project will adopt some important measures such as the improvement of hospital management, adults education, and breastfeeding promotion campaigns. Hopefully, the declining breastfeeding trend will be counteracted.

References

1   WHO: Contemporary patterns of breast-feeding: Report on the WHO Collaborative Study on Breast-Feeding. WHO, Geneva, 1981.
2   National Coordination Working Group on Breast-Feeding Surveillance: A National Breast-Feeding Survey. Chung Hua I Hsueh Tsa Chih 1987;67:433–437.
3   Chengdu Coordination Working Group of Breast-Feeding Survey: 0–6 months infants feeding practices in Chengdu. J Sichuan Med 1985;6:259–262.
4   Ho ZC: Feeding practices, growth and nutritional status of infants and south China. 1st Int Symp Maternal and Infant Nutrition Guangzhou, 00, pp 85–90.
5   Zhu X, Guo Z, Liu X: The effect of feeding patterns on the increment of weight and height in infants (0–6 months). Acta Nutr Sin 1984;6:189–192.
6   Zhu X, Guo Z, Liu X: The effect of feeding patterns on the increment of weight and height in infants (6–12 months). Acta Nutr Sin 1986;8:178–180.
7   Cheng M, Hu X: Study on the morbidity of infants on different feeding patterns during their first 4 months of life. Chin J Pediatr 1986;24:59–61.
8   Liu LX: A study of the effect of 3 different feeding patterns on the physical growth and morbidity of infants during their first 4 months of life. Proc Child Dev Symp (Beijing) 1984;12:48–52.
9   Wang W, Yin T, Li L, et al.: Studies on the relation between the nutritional status of lactating mothers and milk composition as well as the milk intake and growth of their infants in Beijing. 1: Observations on effects of the nutritional status of lactating mothers on breast-milk output and its protein, fat and lactose contents. Acta Nutr Sin 1987;9:338–342.

10   Ho ZC: A survey of breast-feeding in the rural area of Xin-Hui District. Acta Nutr Sin 1983;5:3–10.
11   Hu BD: A survey of breast-milk cessation of rural lactating mothers of different nutritional status. Proc 5th Natl Congr Nutr China, 1988, p 20.
12   Liu DS, Fu A, Jin Y, et al.: A longitudinal study on the dietary intakes of lactating mothers and its influence on breastfeeding. Acta Nutr Sin 1988;10:297–304.
13   Whitehead RG: Maternal Diet, Breastfeeding Capacity, and Lactational Infertility. The United Nations University, 1983, pp 37–41.
14   Yin T, Liu DS, Li L, et al.: Studies of the relationship between the nutritional status of lactating mothers and milk composition as well as the milk intake and growth of their infants in Beijing. V: Essential inorganic elements and vitamins in human milk. Acta Nutr Sin 1989;11:233–239.
15   Packard VS: Human Milk and Infant Formula. New York, Academic Press, 1982.
16   Guthrie HA: Introductory Nutrition. St. Louis, Mosby, 1983.
17   Jin G, Wang C, Gong J, et al.: Studies on the relation between the nutritional status of lactating mothers and milk composition as well as the milk intake and growth of their infants in Beijing. III: The lipid composition of breast milk. Acta Nutr Sin 1980;10:134–144.
18   Zhao X, Xu Z, Wang Y, et al.: Studies on the relation between the nutritional status of lactating mothers and milk composition as well as the milk intake and growth of their infants in Beijing. IV: The protein and amino acid content of breast milk. Acta Nutr Sin 1989;11:227–232.
19   A UNICEF Situation Analysis: Children and Women of China. Beijing, UNICEF, 1989, p 49.

Dong-Sheng Liu, Institute of Nutrition and Food Hygiene,
Chinese Academy of Preventive Medicine, 29 Nan Wei Road,
Beijing 100 050 (China)

Simopoulos AP, Dutra de Oliveira JE, Desai ID (eds): Behavioral and Metabolic
Aspects of Breastfeeding. World Rev Nutr Diet. Basel, Karger, 1995, vol 78, pp 139–163

....
Breastfeeding Trends in Cuba

Manuel Amador[a], *Luis C. Silva*[b], *Graciela Uriburu*[c], *Marta Otaduy*[d],
Francisco Valdés[e]

[a] Institute of Nutrition and Food Hygiene;
[b] Higher Institute of Medical Sciences;
[c] World Food Program;
[d] Faculty of Public Health, and
[e] Ministry of Public Health, Havana, Cuba

Contents

Background

    Breastfeeding is currently considered to be the optimal practice for ensuring the healthy and normal growth of infants, and contributes to the full expression of the genetic potential of an individual's development. For these reasons, the WHO and UNICEF have recommended that every infant should be exclusively breast-fed at least up to the 4th or 6th month of life, and that breastfeeding, complemented with other foods be prolonged at least up to 1 year of age [1]. This has also been recommended by the American Academy of Pediatrics [2].

The advantages of breastfeeding include not only the nutritional adequacy of human milk, but also other important aspects such as the prevention of gastrointestinal and respiratory illnesses, infections and certain immunologic disorders. Growing emphasis is also given nowadays to the role of breastfeeding in the reduction of the risk of certain chronic diseases in adulthood [3].

In the present century, a decline in breastfeeding has been observed in the industrialized countries of North America, Europe and Australia, a trend which has spread rapidly to many less-developed countries in the last three decades, involving poor families as well as the affluent, especially in semiurban areas [4–7].

Reports from many developing countries suggest that this change is occurring with increasing rapidity and with disastrous consequences for the nutrition of young children, with a rising incidence of marasmus and weanling diarrhea in early infancy causing high mortality, an increased risk of permanent ill effects among survivors and a heavy financial burden on the health care services [8–11].

The decline in breastfeeding in the developing world is particularly dangerous because of the poor socioeconomic environmental conditions which prevail there [12]. Complex, interlocking social pressures either preventing a woman from breastfeeding or interfering with her ability to do so may be responsible for the sudden change in the breastfeeding pattern of a community. Women's participation in the work force without adequate maternity leave, the promotion of infant foods in the absence of a solid background policy in favor of breastfeeding and other expressions of new life-styles may be factors accounting for this trend.

In contrast, a tendency toward higher prevalence and longer duration of breastfeeding had been reported in the last 20 years in most industrialized countries [13–21]. In the United States, this favorable trend has been identified in various ethnic groups, particularly among Mexican-Americans [22, 23], but a new decline in overall prevalence and duration at the end of the 1980s has also been described [24].

A great deal of concern has been generated among governments and international institutions such as the WHO and UNICEF about the declining trend in breastfeeding practices in the Third World. In a joint effort, in 1979 the WHO and UNICEF [1] urged the governments from all parts of the world as well as nongovernmental organizations to design and implement programs or strategies to promote breastfeeding, and introduce policies to control the marketing of breast milk substitutes.

In Latin America and the Caribbean, these programs were designed to provide actions for the promotion of breastfeeding starting from institutions and health personnel and going out into the community. In most countries, the initial steps of these programs or strategies included training of personnel working in health institutions or institutions; coordinating with health ministries, and the institutionalization of breastfeeding practices in prenatal, maternity and pediatric services or clinics and hospitals. In addition, these programs brought about significant changes in the standards and norms of maternity services towards directly or indirectly promoting breastfeeding. These new standards included 'rooming in', immediate initiation of breastfeeding, restriction of bottle feeding and the regulation of breast milk substitutes.

Socioeconomic measures such as maternity leave and a 'breastfeeding hour' at the work place are now common practices in many countries, but there are many others in which a woman's right to breastfeed is not yet protected. In 1989, the joint WHO/UNICEF statement on 'Protecting, Promoting and Supporting Breast Feeding: The Special Role of Maternity Services' set out the 'Ten Steps for Successful Breast Feeding' [25].

In Florence, Italy, participants at the WHO/UNICEF Policy Makers Meeting on 'Breast Feeding in the 90s: A Global Initiative', cosponsored by the Agency for International Development of the United States of America (USAID) and the Swedish International Development Authority (SIDA), adopted the 'Innocenti Declaration on the Protection, Promotion and Support of Breast Feeding', recommending the adoption of measures needed to ensure adequate nutrition of the mother and her family as a requirement for optimum health, and the establishment of policies, objectives and a plan of action to promote breastfeeding in the 1990s as a part of maternal and child health programs at the national level [26].

In September 1990, the World Summit for Children Conference approved the 'World Declaration on the Survival, Protection and Development of Children' as well as the plan of action to reduce deaths, which sets as a specific target: 'the empowerment of all women to breastfeed their children exclusively for four to six months and to continue breastfeeding with complementary foods, well into the second year'.

In June 1991, the concept of the 'Baby Friendly Hospital' was launched at an IPA/UNICEF/WHO international breastfeeding meeting. The objectives of the initiative are to mobilize the involvement of maternity hospitals and wards as well as hospital personnel and health professionals towards technically supporting breastfeeding and to create a women's demand for hospitals to be optimally supportive of mothers who wish to breastfeed. The framework for identifying institutions and facilities as 'Baby Friendly' or better, 'Mother and Baby Friendly', is the joint WHO/UNICEF statement [25]. Twelve countries were chosen to set up this initiative and their leading maternity hospitals were the first to be designated 'Baby Friendly Hospitals'.

In Cuba, since the establishment of the Acute Diarrheal Diseases Control Program in 1962, particular emphasis has been given to promoting breastfeeding as part of a comprehensive strategy for health promotion and illness prevention in early life. However, little information about infant feeding patterns existed prior to the 1970s. In a cohort study of 4,272 newborns up to 7 months of age carried out in 1973, Moreno and Rubí [27] found that while a large proportion of mothers started breastfeeding their infants after delivery, the number decreased rapidly thereafter. From a prevalence of 89.8% at discharge from the maternity hospitals, the proportion of exclusively breastfed infants dropped to 35.7% at 4 months of age.

The National Mother and Child Care Program was established in 1980. It included a group of actions to be taken for increasing the prevalence and duration of breastfeeding. In 1984, a new model of primary health care was introduced. According to this model, a physician and a nurse take care of 600 or 700 persons and those attending day care centers, schools and other children's institutions or who work in large centers within a given health area. This activity represents another step towards improving the quality of ambulatory health care, and the new model – the family doctor – gradually substituted the previous one, based on four specializations – pediatrics, internal medicine, obstetrics and dentistry – at the community polyclinic. In 1991, 67.8% of the Cuban population was within the family doctor model of care.

In 1988, the country defined its strategies and actions for preserving and improving the nutritional status of the population through the National Food and Nutrition Program (NFNP). The main goals of the NFNP were : (1) to develop a nationwide multisector campaign for better food habits; (2) to improve the National Food and Surveillance System (SISVAN); (3) to improve the Mother and Child Care Program, and (4) to improve the primary health care network, extending the family doctor model of care to all [28].

To define, design and carry out actions which would allow the attainment of the indicators recommended by the international organizations for breastfeeding practice, it was necessary to know the magnitude of the problems at the national level, since up to the end of the 1980s, only partial information was available. In 1989, a pilot study was designed and conducted prior to a nationwide survey, the results of which have already been published [29, 30]. Finally, the National Survey of breastfeeding and feeding practices of Cuban infants took place in 1990.

The National Survey

The National Survey of breastfeeding and feeding practices of Cuban infants was carried out from February to April 1990. To study the prevalence and duration of breastfeeding, a *status quo* method to obtain relevant information was employed: this is a cross-sectional study, in which inquiry about each subject's status concerning different items is made.

The survey was addressed to the mothers of infants aged 0–364 days. For each infant, date of birth – for obtaining the age in days – as a quantitative variable, and the status

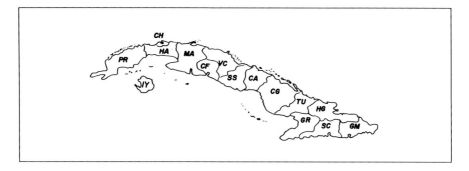

Fig. 1. Strata of the National Survey. Eastern provinces: GM = Guantánamo; SC = Santiago; GR = Granma; HG = Holguín; TU = Las Tunas; CG = Camagüey. Central provinces: CA = Ciego de Avila; SS = Sancti Spíritus; VC = Villa Clara; CF = Cienfuegos. Western provinces: MA = Matanzas; HA = Havana; CH = City of Havana; PR = Pinar de Río; IY = Isle of Youth.

concerning 11 different items were registered. The latter included: the infant's sex; the mother's age, educational level, parity and employment; if the infant was breastfed or not (a qualitative dichotomous variable); if breastfeeding was exclusive or not; if not exclusive, what milk formula was used; if foods other than milk were already introduced and if so which ones.

Information about the model of health care (family doctor or community polyclinic) and type of settlement (urban or rural) were also recorded for each individual.

Description of the Sample

An equiprobabilistic sample comprising all infants (aged 0–364 days) in Cuba was obtained. Sampling was stratified in two stages: the 15 strata were the fourteen provinces and the Special Municipality of Isle of Youth, into which the country is currently divided politically and administratively (fig. 1). In the first stage, health areas within each stratum were chosen; in the second, basic units (family doctor's clinics or infant's health surveillance clinics at the polyclinics) were randomly selected. A group of 92 municipalities, from the 169 existing in the country, were involved in the study. The sample was originally planned to involve a total of 7,000 infants; the final size was 6,688, i.e. only a 0.5% rate of 'no answer'. Table 1 shows the sample distribution according to provinces and urban or rural location.

Techniques and Procedures

The mathematical model of logistic regression, which is based on the observation of several variables and pretends to quantify their effects upon the probability of occurrence of a given event was employed. The analytical form of the curve is:

$$P(x) = P(Y = 1 \mid x) = \frac{1}{1 + e^{(-\alpha - \beta x)}}.$$

Table 1. Sample distribution according to provinces and location

Stratum	Total	Urban	Rural
Pinar del Río	240	166	74
Havana	652	496	156
City of Havana	1,373	1,373	0
Isle of Youth	27	17	10
Matanzas	424	335	89
Cienfuegos	233	192	41
Villa Clara	604	339	265
Sancti Spíritus	251	184	67
Ciego de Avila	156	123	33
Camagüey	379	302	77
Las Tunas	326	69	257
Holguín	431	254	177
Granma	387	140	247
Santiago	927	651	276
Guantánamo	278	246	32
Total	6,688	4,887	1,801

When a dichotomous variable 'Y' (with values 1 or 0) and an 'explicative' variable x are given, the logistic regression expresses the probability of occurrence of a given event ($Y = 1$) for each value of x. Values of α and β are two constants which are estimated through an adequate computational procedure [31]. When the coefficients α and β which define the function are estimated, the goodness of fit can be assessed [31]. The geometric form of the curve is shown in figure 2, where the decreasing pattern of prevalence with age can be observed.

The *desertion index (DI)* from exclusive breastfeeding (EBF) in the first 3 months was calculated. EBF was defined as the practice of feeding only mother's milk.

$$DI = \frac{P_0 - P_{90}}{P_0} \times 100,$$

where P_0 is the prevalence of EBF at birth and P_{90} is the prevalence at 90 days. DI expresses the percentage of EBF infants that had stopped the practice at 90 days.

The *cumulative indices* for the first 4 months of life for each mode of milk feeding were: (1) EBF; (2) combined or mixed feeding (breast plus bottle, MF), and (3) bottle or artificial feeding (no breast feeding, AF). For each 4-month period, a percent distribution of the infants belonging to each one of the above groups can be obtained. The infants usually start in group 1, then shift into group 2 and, finally, go into group 3; thus, the distribution changes with age. The aim is to quantify the dynamics of distribution for each 4-month period.

Figure 2 shows the prevalence curves for EBF and total (overall) breastfeeding (TBF = EBF + MF). A cross section at 120 days has been made and the three areas (A, B and C) which these two curves determine in the rectangle with base 0–120 and height as 100

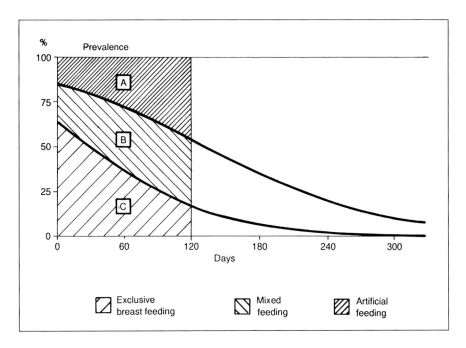

%
Prevalence
100

75

50

25

0
0   60   120   180   240   300

Days

Exclusive breast feeding   Mixed feeding   Artificial feeding

Fig. 2. National prevalence curves of the different modes of milk feeding, Cuba 1990 [32].

are calculated as follows. The area between x_1 and x_2 (in this case $x_1 = 0$ and $x_2 = 120$) under the curve is given by the integral:

$$I = \int_{x_1}^{x_2} P(x)\, dx,$$

with I given by the formula:

$$I = \frac{1}{\beta} \ln\left[\frac{1 - P(x_1)}{1 - P(x_2)}\right].$$

Knowing that $A + B + C = 12{,}000$ (fig. 2) and employing I for both functions, the cumulative indices can be calculated as follows:

$$EBFCI = \frac{C}{12{,}000} \times 100,$$

$$MFCI = \frac{B}{12{,}000} \times 100,$$

and

$$AFCI = \frac{A}{12{,}000} \times 100$$

These three indices were calculated for each province and for the entire country.

Breastfeeding in Cuba

Table 2. Prevalence of different modes of milk feeding of infants at different ages in Cuba, 1990

Age, days	EBF, %	MF, %	AF, %
0	62.7	21.5	15.8
15	56.1	25.3	18.6
30	49.3	28.9	21.8
90	24.5	38.0	37.5
120	15.7	36.7	47.6
180	5.9	27.5	66.6
364	0.2	4.0	95.8

Prevalence and Duration of Breastfeeding

The national prevalence curves of overall TBF and EBF are shown in figure 2. At birth, the prevalence of TBF is 84.2% and that of EBF is only 62.7%. The slope of the former is the gentler, and only a small proportion of infants are still fed breast milk combined with the bottle at 365 days. At 200 days, 6% of the infants are still EBF [32]. Table 2 shows the prevalence of the different modes of feeding at different ages. These figures differ very little from those reported by Moreno and Rubí [27] 20 years ago in a nationwide sample, but are higher than those reported in selected samples or infants from the City of Havana studied in the last decade [33, 34]. All this suggests that very little progress, if any, has been attained concerning breastfeeding practices in this period.

Regional differences in breastfeeding prevalence within a country have been reported. In the United States, the practice is more frequent in the west and less popular in the midwest and east [18, 19, 21]. In Cuba, the curves obtained for each province allow a comparison of the prevalences according to age. For EBF, some regional tendencies could be observed. Eastern provinces, which are also the less developed, show a pattern of high initial prevalence (at birth) with a gradual descending slope and long duration (fig. 3). Central provinces show a pattern of lower prevalence at birth and rapid drop after 3 months with short duration (fig. 4), while western provinces, which include the City of Havana, show diverse prevalences at birth (highest in Pinar del Río and lowest in the capital city), with a rapid drop at 3 months and short duration of EBF (fig. 5).

For all infants who are breastfed (EBF + MF), the curves show differences according to the different regions, the highest prevalences being observed for the eastern provinces. Las Tunas shows the highest prevalence at birth which remains high at 365 days (25.3%, fig. 3). Central and western provinces show similar patterns in their TBF curves, with a more pronounced slope after 3 months; though initial prevalences are somewhat lower in the west (fig. 4, 5).

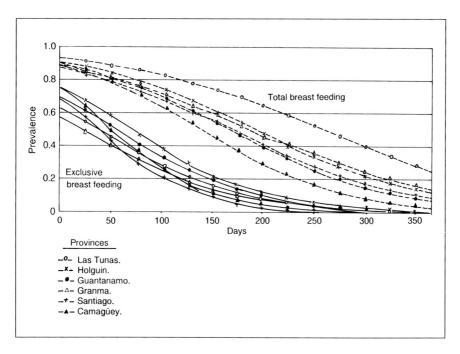

Fig. 3. Patterns of TBF and EBF in eastern provinces of Cuba, 1990. High initial prevalences with gentle descending slopes and long duration are characteristic.

In almost all the provinces, a low percentage of EBF was observed beyond 200 days. This is particularly evident in Las Tunas, Granma, Holguín and Guantánamo which show, however, high prevalences of TBF at 365 days, pointing to the early introduction of bottle feeding as a complement to breastfeeding in those provinces. The low prevalence of EBF at birth in some provinces like Ciego de Avila (44.7%) and Cienfuegos (46.0%) seems to be related to organizational practices in their maternity services which discourage the practice of breastfeeding.

Figure 6 shows the differences in the prevalence of EBF at birth and at 3 months of age, by province and in the whole country. A drop is prominent in all the provinces, especially in three of the western ones, including the City of Havana. In table 3, the provinces have been ordered according to the magnitude of the DI. Low prevalences at birth are probably strongly related to organizational procedures in the maternity services where 99.8% of the deliveries occurred in Cuba in 1990. High DIs point to other aspects related to primary health care organization but could also be related to the cultural patterns of the population, since a tendency to lower DIs in the eastern provinces (less developed and with a more rural population) has been observed.

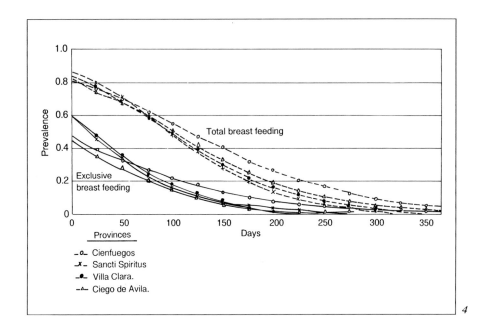

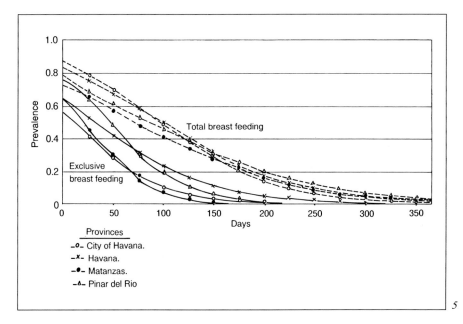

Fig. 4. Patterns of TBF and EBF in central provinces of Cuba, 1990. Lower initial prevalences and a rapid drop after 3 months, with short duration, can be observed.

Fig. 5. Patterns of TBF and EBF in western Cuban provinces, 1990. Initial prevalences are diverse, but duration is short after a rapid drop at 3 months.

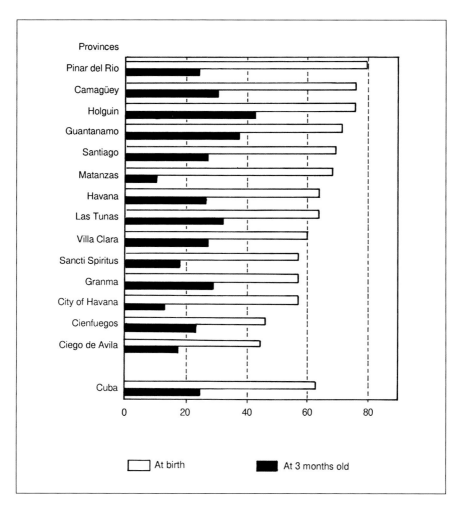

Provinces
- Pinar del Rio
- Camagüey
- Holguin
- Guantanamo
- Santiago
- Matanzas
- Havana
- Las Tunas
- Villa Clara
- Sancti Spiritus
- Granma
- City of Havana
- Cienfuegos
- Ciego de Avila

- Cuba

0   20   40   60   80

☐ At birth   ■ At 3 months old

Fig. 6. Prevalences of EBF at birth and at 3 months of age, Cuba 1990 [32]. A drop is prominent in all the provinces, especially in three of the western ones.

Figure 7 shows the cumulative indices by province from birth to 120 days. Particularly striking is the magnitude of the AFCI, which is very high in Matanzas (46.0%) and Pinar del Río (41.9%), in spite of the latter having the highest prevalence of EBF at birth. As expected, the highest cumulative indices for breastfeeding (EBFCI + MFCI) were shown by the eastern provinces. The trend in artificial feeding (AFCI) is for an increase from east to west, corresponding to the drop in TBF observed at 3 months. This finding is probably associated with urban cultural patterns.

Table 3. Desertion from EBF at 3 months of age by province, Cuba 1990 [32]

Province	EBF at birth, %	EBF at 3 months, %	Differ-ence	DI %
Matanzas	68.0	10.1	57.9	85.1
City of Havana	57.1	13.1	44.0	77.1
Pinar del Río	79.4	24.1	55.3	69.6
Sancti Spíritus	55.1	17.7	39.4	69.0
Ciego de Avila	44.7	17.2	27.5	61.5
Santiago	69.0	26.8	42.2	61.2
Camagüey	75.4	30.1	45.3	60.1
Havana	63.6	26.2	37.4	58.8
Villa Clara	59.9	26.8	33.1	55.3
Granma	57.1	28.7	28.4	49.7
Cienfuegos	46.0	23.3	22.7	49.3
Las Tunas	63.1	32.0	31.1	49.3
Guantánamo	71.1	36.9	34.2	48.1
Holguín	75.4	42.4	33.0	43.8
Cuba	62.7	24.5	38.2	60.9

Table 4. Infants with EBF and those fed only breast milk (OBM) according to age, Cuba 1990 [32]

Age group days	Total	EBF		OBM	
		n	%	n	%
<15	312	260	83.3	233	74.7
16–30	370	227	61.4	75	20.3
31–60	764	298	39.0	72	9.4
61–90	701	149	21.3	17	2.4
91–120	677	106	15.7	10	1.5
121–364	3,864	190	4.9	4	0.1
Total	6,688	1,230		411	

In table 4 the data corresponding to the infants fed only breast milk (OBM) compared to those EBF are summarized. The differences between these two groups is that OBM infants are those who receive no food other than breast milk, whereas EBF infants receive milk only from the mother's breast, but with the possible addition of other foods. The reason for making such a distinction is that

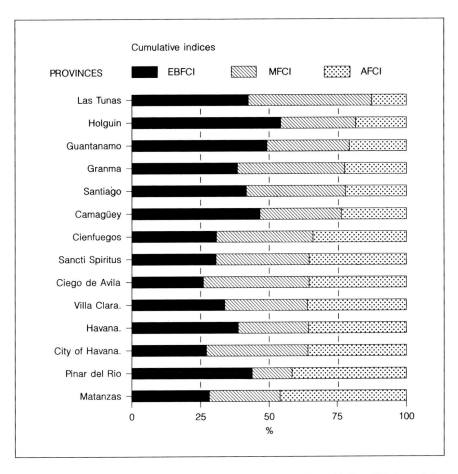

Fig. 7. Cumulative indices of milk feeding by province, from birth to 120 days, Cuba 1990 [32]. The highest figures for cumulative breastfeeding can be observed in the six eastern provinces.

in Cuba, mothers introduce fruit juices and purees and some solid foods very early in the first 4 months of life, when breast milk should be the sole food for the infants [27, 34]. Of 312 infants up to 15 days old, 83.3% were breastfed, but only 74.7% received breast milk alone (OBM). These proportions change very quickly in the second half of the first month of life when the percentages fall to 61.5 and 20.3%, respectively. By 120 days of age, 94.7% of the infants were fed fruit juices and for 67% meat had already been introduced [35].

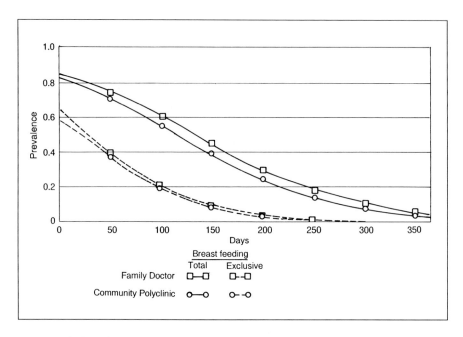

Fig. 8. Prevalence curves for TBF and EBF for Cuban infants under two different models of primary health care, 1990. Note the slight advantage in initial prevalence and longer duration of both modes in infants under the family doctor's model of care.

The Family Doctor and Mode of Feeding: Differences between Urban and Rural Areas

The effect of the model of health care with the family doctor and the differences between urban and rural areas in the prevalence of breastfeeding were also investigated [36]. The methodologic procedure employed was, as before, the mathematical model of univariate logistic regression. The analysis was carried out for each of the four subpopulations of interest resulting from the combination of the categories of both axes: location (urban-rural) and model of health care (family doctor-community polyclinic). The logistic function was measured twice: for first curve, it was considered an explicative variable if the infant was or was not breastfed; for the second, it was considered if the infant was exclusively breastfed or not. From a total of 6,688 infants, 4,295 were in the care of the community polyclinic and 2,393 were under a family doctor.

Figure 8 shows the prevalence curves for overall (total) breastfeeding (EBF + MF) and EBF for infants under the family doctor or attending a com-

Table 5. EBFCI and AFCI according to the model of health care and location

Model of health care	EBFCI		AFCI	
	urban	rural	urban	rural
Family doctor	32.8	50.7	35.0	19.8
Community polyclinic	34.1	44.6	31.3	23.5

munity polyclinic. For TBF, both curves are very similar, showing a gentle slope and running almost parallel from birth to 365 days, but with the curve for the family doctor always running above that for the community polyclinic. For EBF, the family doctor curve starts slightly above the community polyclinic curve, but both continue almost together to the end, showing a rapid descending slope at 90 days. Though at the national level the differences were not significant, the pilot study carried out in four polyclinics of the City of Havana showed a median for the duration of TBF of 107 days for family doctor infants compared to 74 days for those under community polyclinics. At birth, there were no differences in prevalences, but they became evident at 90, 180 and 360 days, suggesting that the model of health care influences the duration of breastfeeding more than its initiation [29].

The EBFCI and AFCI according to location and model of infant health care appear in table 5. From the analysis of these data, it is evident that EBF corresponds to approximately one third of the infants in the first 4 months of life in urban areas, irrespective of the model of health care, but is significantly higher in rural areas under the family doctor, where it is achieved by more than one half of the infants. Conversely, AFCI shows lower figures in rural areas especially those under a family doctor. These results are consistent with various reports which indicate that mothers from rural areas breastfeed more and for longer than urban ones [4, 37–39]. The decline in breastfeeding observed in many countries occurs mainly in urban areas [4].

Mother's Characteristics and Breastfeeding

Family and maternal characteristics are associated with patterns of infant feeding and these associations are diverse across countries [39]. In the National Survey on breastfeeding and feeding practices in Cuba, four maternal characteristics were recorded: age, parity, educational level and occupation.

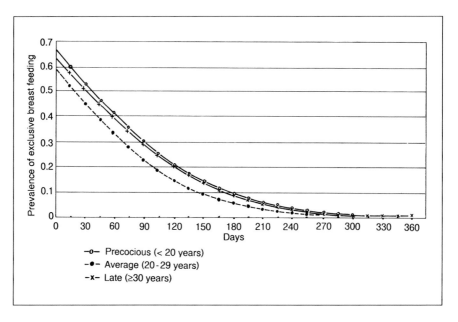

Fig. 9. Prevalence of EBF according to mother's age, Cuba 1990. Precocious (adolescent) mothers breastfeed more and for longer.

Exclusive Breastfeeding and Mother's Age

For this purpose, the sample was classified into three groups according to the mother's age: precocious, <20 years (1,337 mothers); average, 20–29 years (4,348 mothers), and late, ≥30 years (1,003 mothers). Prevalence curves appear in figure 9. Precocious mothers show the highest prevalence at birth while later, the curve parallels and closely approaches that of late mothers. Average mothers show the lowest prevalences throughout, and the lowest EBFCI at 4 months (33.8%). Conversely, the highest indices are found among the precocious mothers (41.7%) followed by the late ones (40.5%). The higher prevalence of EBF among precocious mothers is interesting since an opposite trend might have been expected. Social and familial background in adolescent mothers may influence patterns of infant care and feeding in a negative sense since pregnancies in teenagers are more likely to be undesirable to occur in unmarried girls with inadequate economic and emotional support. Nevertheless, studies with teenage mothers show that most of them express positive attitudes toward the pregnancy and the infant [40] and no relationship has been found between adolescent maternal age and mothering behaviors [41]. Successful lactation can be attained through active counseling before and after delivery [42].

In several studies, the mother's age has correlated positively with the length of breastfeeding [43, 44]. Valdés-Díaz et al. [45] in a survey in Havana found that

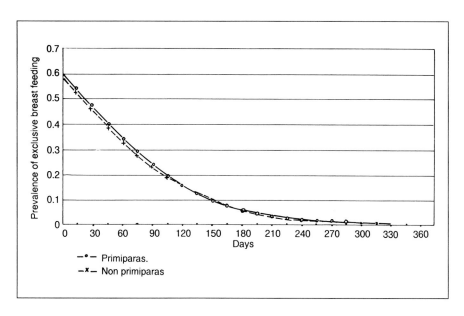

Fig. 10. Prevalence of EBF according to mother's parity, Cuba 1990. No differences in initiation and duration can be observed.

mothers ≥ 30 years old breastfed more and for longer. With 220 mothers, Grossman et al. [46] found that women who chose to breastfeed were likely to be older, but other authors have found no age differences [23, 47], or have observed that the initiation but not duration decreased with mothers > 30 years [33, 44, 48]. On the other hand, several studies have shown that women < 20 years of age are less likely to breastfeed their infants [21, 45, 49]. In Cuba, there are reports which are consistent with the findings of the National Survey: in a health area of Cienfuegos, Cuéllar et al. [50] found that mothers < 20 years were more likely to breastfeed for > 3 months, and in the follow-up of 1,483 newborns from two health areas of the city of Havana, Valdés et al. [33] found that the initiation of EBF was 90.2% in adolescent mothers compared to 70.1% in mothers > 35 years old.

Mother's Parity and Exclusive Breastfeeding

Mothers were grouped into two categories, according to their parity: *primiparas*, if they were having the first delivery (n = 3,746) and *nonprimiparas* if they had had previous deliveries (n = 2,942). No difference in prevalence according to parity was observed and the curves were practically superimposed along the year, especially after 2 months of age (fig. 10). The cumulative indices (EBFCI) show no significant difference: 35.5% for primiparas and 34.9% for nonprimiparas, at 4 months of age.

Differences between the two groups are based upon the existence or not of a previous delivery, a clear distinction, and not on the basis of the previous experience in breastfeeding, since no retrospective data were collected in the National Survey. Several reports indicate that the feeding decision made by mothers with more than one child is largely determined by their experience in feeding the previous children [51] and that mothers more experienced in breast-feeding breastfeed longer [46, 52]. Nevertheless, results from two large studies carried out in Canada [20, 53] and a survey of 1,954 mothers in France [47] showed that the initiation of breastfeeding was more common among primiparas, whereas it had a greater duration among multiparas [20, 44].

When considering parity as a variable, it is necessary to take into account its association with other factors. Maternal age and parity are highly correlated, their separate effects on breastfeeding patterns being difficult to distinguish. In the WHO Collaborative Study on Breast-Feeding carried out in nine different countries, it was possible to separate both effects only in Sweden, where the prevalence of breastfeeding was higher among mothers with first babies and among older mothers [39]. Data from national survey in the USA conducted by the NCHS in 1981 showed that parity was negatively related to breastfeeding [44].

Parity is, of course, associated with the experience of breastfeeding which could be good, encouraging the mother to breastfeed her next child, or negative, discouraging her from doing so. The results of the National Survey show, therefore, that having previous children does not affect infant feeding behavior.

Educational Level and Exclusive Breastfeeding

Mothers were grouped into three categories according to their educational level: elementary school, up to 6 years (971 mothers); junior high school, from 7 to 9 years (3,076 mothers), 2nd senior high school + university, ≥ 10 years (2,641 mothers). Mothers in the elementary school group breastfed more and for longer. They show a higher prevalence at birth too. The lowest prevalences were found among mothers with the highest educational levels. Though they show higher prevalence at birth than the group with junior high school education, the slope of the curve falls more rapidly, and is significantly lower at 60 days of age (fig. 11).

At 4 months of age, the highest cumulative indices (EBFCI) are found in mothers with elementary school (43.4%), followed by those with junior high school (36.4%) and senior high school and university (31.7%) education. Patterns of breastfeeding according to the mother's educational level seem to be different in the developed countries than in developing ones. Various studies carried out in the United States [17, 44, 46, 54], Scandinavian countries [16] and France [47] have shown that more educated women are more likely to breastfeed. This trend has been observed among different ethnic groups in the United States too [23, 55]. In the WHO Collaborative Study on Breast-Feeding [39], it was observed that in

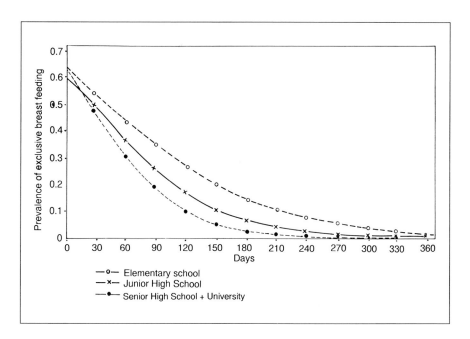

Fig. 11. Prevalence of EBF according to mother's educational level, Cuba 1990. Mothers with elementary school level education feed more and for longer. Initiation of breastfeeding is more frequent among them too.

Sweden, at 3 months postpartum, breastfeeding was more common among mothers with a higher educational level, while in India and the Philippines the trends observed were in the opposite direction. In Latin American countries, studies have shown that the duration of breastfeeding is longer among less-educated women [48, 56, 57]. Since educational level is commonly associated with economic status, in some cases it is not easy to separate both effects [48, 57]. In a large study in the metropolitan area of Washington DC, it was observed that education, not income, was the important predictor of breastfeeding in white women, whereas for black women it depends on income as well [49]. In Cuba, the National Survey shows a trend that resembles that of a developing country, though some local reports point to a higher prevalence and duration among more-educated women [45, 50, 58].

Mother's Occupation and Exclusive Breastfeeding
Employed mothers (n = 3,612) show significantly higher breastfeeding prevalences at birth, but the slope immediately after discharge from maternity wards is more pronounced, and there are no infants with EBF after 240 days of life

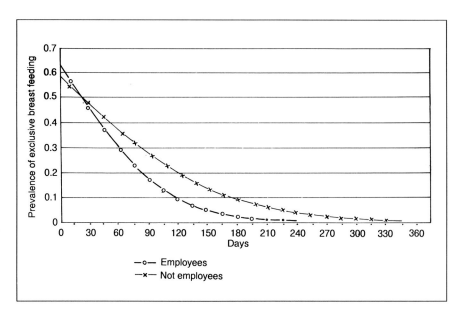

Fig. 12. Prevalence of EBF according to mother's occupation, Cuba 1990. Employed mothers initiate breastfeeding more often, but the descending slope is more pronounced especially after 3 months, compared with unemployed mothers.

in this group. For unemployed mothers (n = 3,076), the slope is gentler and the prevalence significantly higher than for employed mothers at 3 months, and there is a small group of infants who are breastfed beyond 240 days (fig. 12). Employed mothers show lower cumulative indices (EBFCI) at 4 months (31.7%) than the unemployed mothers (37.9%). This significant difference between employed and unemployed mothers occurs in spite of the existence in Cuba of legislation which allows for 6 weeks maternity leave before delivery and 12 weeks after, with full salary. Recently, the benefits of the law were extended up to 6 months with 60% of the salary and up to 12 months without payment, but with job preservation. Hence, the economic implications of motherhood are reduced and social and biological aspects are supported. Maternal employment has a large impact on the duration of EBF: in the Ross National Surveys, Ryan et al. [24], reported that by the fourth month after delivery, employed mothers had a higher probability of initiating MF (43%) compared with mothers who were not in the work force (19%). In a study of 1,900 mothers participating in the Johns Hopkins University Infant Feeding Study in Baltimore (Md., USA), less than one half of employed mothers were still breastfeeding within the first 3 months postpartum, whereas two thirds of those who were not employed were still breastfeeding [59].

The decline in breastfeeding by the 3rd month postpartum has been associated with the end of maternity leave and the need of the mothers to prepare for reassuming their posts. In Cuba, breastfeeding has been encouraged by government policy and by mass organizations, especially the Cuban Women's Federation, but facilities have also been given to employed mothers to keep their infants in day care centers as early as 45 days of age. This practice decreased the likelihood that employed mothers would start breastfeeding or extend it beyond the 2nd month of life, and probably incited the practice of early introduction of solid foods which has characterized infant feeding patterns in the last 20 years [27, 34, 35]. The extension of maternity leave now makes the early admission of infants to day care centers unnecessary and gives mothers the opportunity to stay with them for longer and to start weaning properly.

The mother's occupation is undoubtedly associated with many other factors which may be playing a role in the trends observed in the prevalence and duration of breastfeeding. A complex number of social, economic, individual and familial variables which were not included in the National Survey are surely interacting, and could explain why bottle formula is introduced to the breastfed infant by a mother who is protected by the law.

Summary of the National Survey Results

The conclusions of the National Survey on breastfeeding and feeding practices of Cuban infants can be summarized as follows:

(1) The pattern of infant feeding in Cuba differs substantially from current international recommendations.

(2) An initial low prevalence of EBF was observed. This finding was probably related to organizational practices in the maternity services which interfere with the initiation of breastfeeding.

(3) The rapid fall in breastfeeding prevalence during the first 3 months of life probably points to poor utilization of the well-developed public health network for educational actions.

(4) Encouraging results in prevalence and duration of breastfeeding in rural areas under the family doctor model of care were observed.

(5) Mothers with a low educational level, <20 or $\geqslant 30$ years of age, and unemployed were more likely to breastfeed. Parity did not influence breastfeeding trends.

(6) The early introduction of bottle feeding and solid foods was observed, which is probably associated with cultural patterns that overestimate the nutritional value of some foods.

Strategies and Actions for Changing Breastfeeding Trends in the 1990s

Starting from the third quarter of 1991, and with the objective of achieving a substantial change in infant feeding patterns by the year 1995, a group of strategies and actions have been undertaken.

December 1991 saw the start of the National Program of Action for reaching the goals of the World Declaration on the Survival, Protection and Development of Children, approved at the World Summit for Children Conference in September 1990. In this program, aspects related with infant feeding were included with three main goals:

(1) To increase breastfeeding initiation to 90% by 1995 and to 95% by the year 2000.

(2) To increase the EBF prevalence at 4 months of age to 50% by 1993, to 70% by 1995 and to 80% by the year 2000.

(3) To maintain breastfeeding complemented with other foods at 6 months of age, reaching a prevalence of 60% by 1995 and 75% by the year 2000.

The strategies and main actions for achieving these goals were established and included in the Program [60] and appear in the document 'Objectives, design and guidelines for increases in the population's health, 1992–2000' issued by the Cuban Ministry of Public Health [61].

Breastfeeding has been integrated into some program components such as diarrheal disease control, family planning and nutrition. Organizational procedures recommended in the 'Ten Steps for Successful Breast Feeding' were put into practice in the maternity services, where 99.8% of all the nation's deliveries occur. The continuity of educational activities for the promotion of breastfeeding during the prenatal period, at the moment of delivery and on the occasion of the monthly follow-up is guaranteed. The Cuban Women's Federation is organizing local support groups led by volunteers which promote breastfeeding and contribute to the mother's information and acquisition of skills. The collaboration of health workers with mass ogranizations at the local level complements the educational programs which are developed in day care centers and schools (elementary and secondary) with the aim of developing educational and cultural patterns, not only among girls but also among boys, favorable to the practice of breastfeeding. In August 1992, a workshop on the 'Baby Friendly Hospital Initiative' sponsored by UNICEF and the Cuban Ministry of Public Health was held in Havana. Six large maternity hospitals from different regions of the country were initially included in this program and another ten joined during 1993. By the end of 1994, the Second National Survey will be carried out in order to make an initial evaluation of the impact of the interventions derived from the above-mentioned National Program of Action. The successful application of these interventions will be decisive for maintaining the high standards which characterize infant health in Cuba.

References

1   WHO/UNICEF: Joint WHO/UNICEF Meeting on Infant and Young Child Feeding: State-
     ment and Recommendations. Geneva, World Health Organization, 1979.
2   American Academy of Pediatrics Committee on Nutrition: Prudent lifestyle for children: Dietary
     fat and cholesterol. Pediatrics 1986;78:521–525.
3   Cunningham AS, Jelliffe DB, Jelliffe EFP: Breast-feeding and health in the 1980s: A global
     epidemiologic review. J Pediatr 1991;118:1–8.
4   Jelliffe DB, Jelliffe EFP: Human Milk in the Modern World. Oxford, Oxford University Press,
     1978, pp 182–299.
5   Soysa PE: The advantages of breast feeding: A developing country point of view. Assign Child
     1981;55/56:25–40.
6   WHO: The prevalence and duration of breast feeding: A critical review of available information.
     World Health Stat Q 1982;2:92–116.
7   Schmidt BJ, Bertrand WE, Mock NB: Infant feeding practices among the poor in Latin America
     and Portugal. Cour Cent Int Enfance 1985;35:361–373.
8   Chàvez A, Martínez C, Bourges H, et al.: Child nutrition problems during lactation in poor rural
     areas; in Chàvez A, Bourges H, Basta S (eds): Nutrition. Basel, Karger, 1972, vol 2: Prognosis for
     the Undernourished Surviving Child, pp 90–91.
9   Juez G, Díaz S, Casado ME, et al.: Growth pattern of selected urban Chilean infants during
     exclusive breastfeeding. Am J Clin Nutr 1983;38:462–468.
10 Rowland MGM: The weanling's dilemma: Are we making progress? Acta Paediatr Scand Suppl
     1986;323:33–42.
11 Launer LJ, Habicht JP, Kardjati S: Breast feeding protects infants in Indonesia against illness
     and weight loss due to illness. Am J Epidemiol 1990;131:322–331.
12 Popkin B, Bilsborrow R, Akin JJ: Breastfeeding patterns in low-income countries. Science
     1982;218:1088–1093.
13 Sjölin S: Present trends in breastfeeding. Curr Med Res Opin 1976;4(suppl 1):17–22.
14 Lawson JS, Mays CA, Oliver TI: The return to breast feeding. Med J Aust 1978;2:229–230.
15 Verkasalo M: Present trends in breastfeeding in southern Finland. Acta Paediatr Scand 1980;69:
     89–91.
16 Helsing E: Infant feeding practices in northern Europe. Assign Child 1981;55/56:73–89.
17 Martínez GA, Dodd DA, Samartgedes JA: Milk feeding patterns in the United States during the
     first 12 months of life. Pediatrics 1981;68:863–868.
18 Martínez GA: Trends in breast-feeding in the United States; in Report of the Surgeon General's
     Workshop on Breastfeeding and Human Lactation. DDHS publ No HRS-D-MC 84-2, 1984, pp
     18–30.
19 Martínez GA, Krieger FW: 1984 milk feeding patterns in the United States. Pediatrics 1985;76:
     1004–1008.
20 McNally E, Hendricks S, Horowitz I: A look at breast-feeding trends in Canada (1963–1982).
     Can J Public Health 1985;76:101–107.
21 Udall JN, Kilbourne KA: Selected aspects of infant feeding. Nutrition 1988;4:409–415.
22 Forman MR, Fetterly K, Graubard B, et al.: Exclusive breast feeding of newborns among
     married women in the United States: The National Natality Surveys of 1969 and 1980. Am J Clin
     Nutr 1985;42:864–869.
23 John AM, Martorell R: Incidence and duration of breast feeding in Mexican-American infants:
     1970–1982. Am J Clin Nutr 1989;50:868–874.
24 Ryan AS, Rush D, Krieger FW, et al.: Recent declines in breast feeding in the United States: 1984
     through 1989. Pediatrics 1991;88:719–727.
25 WHO/UNICEF: Protecting, Promoting and Supporting Breast Feeding: The Special Role of
     Maternity Services. A Joint WHO/UNICEF Statement. Geneva, World Health Organization,
     1989, pp 1–32.
26 WHO/UNICEF: Policymakers' Meeting on 'Breastfeeding in the 1990s: A Global Initiative'.
     Geneva, World Health Organization, 1990.

27 Moreno O, Rubí A: Estudio de una cohorte de niños desde el nacimiento basta los 7 meses de edad. La Habana, CNICM, Serie Información Temática. 1979, pp 3–73.
28 Amador M, Peña M: Nutrition and health issues in Cuba: Strategies for a developing country. Food Nutr Bull 1991;13:311–317.
29 Silva LC, Baonza I, Amador M: Efecto del médico de la familia en la prevalencia y duración de la lactancia materna. Rev Cubana Pediatr 1989;61:643–653.
30 Silva LC, Baonza I, Amador M: Epidemiología de la lactancia materna: prevalencia y duración. Aten Primaria 1991;8:455–459.
31 Silva LC, Fariñas H: Un programa para regresión logística politómica. Trib Med 1988;3:21–22.
32 Amador M, Silva LC, Uriburu G, et al.: Caracterización de la lactancia materna en Cuba. Food Nutr Bull 1992;14:101–107.
33 Valdés R, Seidedos M, Reyes DM, et al.: Prevalencia y duración de la lactancia materna. Seguimiento de 1483 niños hasta el año de edad. Rev Cubana Pediatr 1989;61:633–642.
34 Amador M, Hermelo MP, Valdés M, et al.: Feeding practices and growth in a healthy population of Cuban infants. Food Nutr Bull 1992;14:108–114.
35 Silva LC, Fuentelsaz C, Amador M: Rasgos caracteristicos de la introducción de alimentos al lactante en Cuba. Bol Of Sanit Panam 1993;114:407–414.
36 Silva LC, Uriburu G, Amador M: Influencia del Médico de la Familia sobre la lactancia materna en Cuba. Rev Saude Publica, in press.
37 Sadre M, Emami E, Donoso G: The changing pattern of malnutrition. Ecol Food Nutr 1971;1:55–57.
38 Berg A: The Nutrition Factor. Washington, Brookings Institution, 1973.
39 WHO: Contemporary Patterns of Breast-Feeding: Report on the WHO Collaborative Study on Breast-Feeding. Geneva, World Health Organization, 1981.
40 Camp BW, Morgan LJ: Child-rearing attitudes and personality characteristics in adolescent mothers: Attitudes toward the infant. J Pediatr Psychol 1984;9:57–63.
41 McAnarney ER, Lawrence RA, Aten MJ, et al.: Adolescent mothers and their infants. Pediatrics 1984;73:358–362.
42 Lipsman S, Dewey KG, Lönnerdal B: Breastfeeding among teenage mothers: Milk composition, infant growth and maternal dietary intake. J Pediatr Gastroenterol Nutr 1985;4:426–434.
43 Rajcoomar V, Wong PC: Breastfeeding and Infant Health in Mauritius. UNICEF Social Statistics Programme No 6, 1986, p 14.
44 Ford K, Labbok M: Who is breast feeding? Implications of associated social and biomedical variables for research on the consequences of method of infant feeding. Am J Clin Nutr 1990;52:451–456.
45 Valdés-Díaz J, Herrera E, Muñoz JR: Lactancia materna y madre adolescente. Rev Cubana Pediatr 1990;62:560–565.
46 Grossman LK, Fitzsimmons SM, Larsen-Alexander JB, et al.: The infant feeding decision in low and upper income women. Clin Pediatr 1990;29:30–36.
47 Nicaud V, Hatton F, Robine JM, et al.: Allaitement maternel: Nature de choix. Arch Fr Pédiatr 1985;42:133–137.
48 Díaz-Roselló JL, Bauzá CA: Estudio epidemiológico sobre prácticas de lactancia materna. IV-Estudio analítico de las asociaciones entre las diversas variables. Arch Pediatr Uruguay 1980;51:3–27.
49 Kurinij N, Shiono PH, Rhoads GG: Breast feeding incidence and duration in black and white women. Pediatrics 1988;81:365–371.
50 Cuéllar MC, Figueroa R, Ramos MJ, et al.: Lactancia materna. Algunos factores que promueven el destete precoz. Rev Cubana Med Gen Integ 1989;5:7–18.
51 Martin J: Infant feeding 1975: Attitudes and Practice in England and Wales. London: Office of Population Censuses and Surveys: Social Survey Division, Her Majesty's Stationery Office, 1978.
52 Lewis NM, Fox HM: Factors associated with breastfeeding duration. Nutr Res 1986;6:1121–1129.
53 Walker DA, Abernathy TJ, Maloff BMK, et al.: Infant feeding practices in Calgary during 1984. J Can Diet Assoc 1987;48:108–112.

54   Martínez GA, Nalezienski JP: 1980 update: The recent trend in breast feeding. Pediatrics 1981;67:260–263.
55   Serdula MK, Cairns KA, Williamson DF, et al.: Correlates of breast-feeding in a low-income population of whites, blacks and Southeast Asians. J Am Diet Assoc 1991;91:41–45.
56   Santos-Torres I, Vásquez-Garibay E, Nápoles-Rodríguez F: Hábitos de lactancia materna en colonias marginales de Guadalajara. Bol Med Hosp Infant Mex 1990;47:318–323.
57   Barros FC, Victora CG, Vaughan JP: Breast feeding and socio-economic status in southern Brazil. Acta Paediatr Scand 1986;75:558–562.
58   Pérez N, Sarmiento G, Muiño MC: La lactancia materna. Factores biosociales que inciden en un área de salud atendida por el Médico de la Familia. Rev Cubana Med Gen Integ 1989;5:178–184.
59   Gielen AC, Faden RR, O'Campo P, et al.: Maternal employment during the early postpartum period: Effects on initiation and continuation of breast feeding. Pediatrics 1991;87:298–305.
60   República de Cuba: Programa Nacional de Acción para el cumplimiento de los acuerdos de la Cumbre Mundial en favor de la Infancia. La Habana, 1991.
61   Ministerio de Salud Pública: Objectivos, propósitos y directrices para incrementar la salud de la población cubana 1992–2000. La Habana, ECIMED, 1992, pp 1–19.

Prof. Manuel Amador, Institute of Nutrition and Food Hygiene (INHA),
Calzada de Infanta 1158, La Habana 10300 (Cuba)

Simopoulos AP, Dutra de Oliveira JE, Desai ID (eds): Behavioral and Metabolic
Aspects of Breastfeeding. World Rev Nutr Diet. Basel, Karger, 1995, vol 78, pp 164–190

....
Breastfeeding Patterns in the Arabian Gulf Countries

Abdulrahman O. Musaiger

Department of Food Sciences and Nutrition, Faculty of Agricultural Sciences,
United Arab Emirates University, Al-Ain, United Arab Emirates

Contents

Introduction

The Arab Gulf countries, namely Bahrain, Kuwait, Oman, Qatar, Saudi Arabia and the United Arab Emirates (UAE) have similar social, cultural, economic and demographic characteristics, with common health problems. These countries have experienced rapid social and economic changes during the past two decades, mostly due to oil revenue. As a result, there has been a great change in lifestyle and health problems. Infant mortality and child morbidity rates have decreased sharply due to dramatic improvements in health services. However, a paradoxical nutrition situation has developed: problems associated with affluence such as obesity, non-insulin-dependent diabetes, hypertension, coronary heart disease, dental caries and cancer are manifested alongside those associated with underdevelopment, such as growth retardation and nutritional anemias [1].

The practice of breastfeeding has declined dramatically in all Arabian Gulf countries. Two decades ago, mothers in these countries continued breastfeeding until the end of the second year of their child's life, and some continued further until they became pregnant again [2]. However, the percentage of mothers who practiced breastfeeding decreased sharply thereafter, especially during the oil boom period (1973–1980). In the 1960s, Sebai [3] studied the state of infant feeding practices in Saudi Arabia and found that breastfeeding was the predominant method up to 2 years of age. In the 1980s, Elias [4] showed that 85% of infants in Saudi Arabia were breastfed immediately after birth; this dropped to 38% at age 6 months and 22% at age 12 months. Similar patterns were also reported in other Gulf countries [5]. Data on breastfeeding patterns in the Gulf show that mixed (breast and bottle) feeding is the norm. The recent National Child Health Surveys which were carried out in six Arabian Gulf States demonstrated that the percentage of mothers who used mixed feeding ranged from 48% in Oman to 64% in the UAE [6].

This paper surveys the current breastfeeding situation, factors influencing it and programs to support breastfeeding in the Arabian Gulf States.

Breastfeeding Practices in the Gulf

Exclusive Breastfeeding

Exclusive breastfeeding (the supply of breast milk without supplementation of any other foods or liquids) not only influences lactational infertility, but also, prevents diarrhea, and provides protection against respiratory infection, eczema and asthma [7]. The duration of exclusive breastfeeding is very short even in countries with the greatest median duration of breastfeeding. It rarely exceeds 1 month in most countries [8].

Information on the duration of exclusive breastfeeding in the Gulf is scarce. A recent survey in Oman showed that 12.9% of mothers exclusively breastfed their infants for 1 week, 3.4% for 2 weeks and 3.7% for 3 weeks. The median duration of exclusive breastfeeding varied according to geographical region, and ranged from 2 to 5 weeks [9]. These percentages were relatively low; however, there are no previous data on exclusive breastfeeding in Oman. It is therefore difficult to assess the decline or increase in the duration of exclusive breastfeeding in this country.

The Omani survey [9] indicated that 30.1% of mothers fed their infants both breast milk and water for 1–3 weeks; the percentage declined to 17.5% for 4–6 weeks, and to 12.5% for 7–9 weeks. There is now scientific evidence to suggest that the exclusively breastfed infant does not require additional water because human milk has a low solute load and osmolality [10].

Initiation of Breastfeeding

Late initiation of breastfeeding has several hazardous effects on mother and infant. It delays mother-infant bonding, increases the risk of hypothermia in the newborn, exposes the infant to diarrheal pathogens, contributes to insufficient milk due to reduced suckling at the breast and may lead to breast engorgement causing great discomfort and putting the mother at risk for breast infections [11]. In addition, suckling immediately after delivery hastens uterine contractions and may reduce the maternal risk of postpartum hemorrhage [12].

Studies in three Arab Gulf states showed that the percentages of mothers who started breastfeeding at the first hour after delivery were 11.5, 28 and 37% in Saudi Arabia, Bahrain and Oman, respectively (table 1). The majority of mothers in Saudi Arabia (83.7%) and Oman (92%) initiated breastfeeding during the first 12 h after delivery, while the percentage was lower among Bahraini mothers (46%).

The late initiation of breastfeeding in the Gulf communities can be attributed to several factors. First, health staff in hospitals and private clinics fail to provide proper guidance. In most hospitals in the Gulf it is a common practice to provide the infant with a formula from the first hours of life. Second, there is a misconception that colostrum is bad for the infant. This belief is still widespread in

---

Musaiger

Table 1. Time of breastfeeding initiation after delivery

Country	Sample size	Time, h				Reference
		< 1 %	1–6 %	7–12 %	12 + %	
Bahrain	499	28.0	15.0	3.0	54.0	16
Oman	1,028	36.9	52.2	2.9	8.1	9
Saudi Arabia	990	11.5	52.6	19.6	16.3	19

some Gulf communities. In Oman, Musaiger [13] found that 27% of mothers believed that colostrum harms the infant. The main reasons given by mothers for this belief were that colostrum causes diarrhea in infants, it is a dirty milk and its color is not acceptable. It is well documented that this milk is beneficial to the infant both nutritionally and immunologically [14]. However, despite these advantages, colostrum is discarded and replaced by infant formula, water, glucose, honey and ritual foods. Finally, mothers (and even health staff) are ignorant about the health benefits of early initiation of breastfeeding for both the mother and her child.

Frequency of Breastfeeding

Frequent suckling of the breast is one of the best methods to stimulate milk secretion. Frequent suckling and emptying of the breasts will also favor better and faster establishment of lactation [15]. Many mothers in the Gulf believe that infants should be breastfed according to a schedule, and therefore limit the number of feedings per day and night. This practice is affected by the mothers' educational level. Educated mothers, in general, are more likely to breastfeed their infants on a schedule. In Bahrain, it was reported that 44% of highly educated mothers (those with secondary degrees and above) breastfed their infant on a schedule, compared to 12 and 25% among mothers who were illiterate or with primary education, respectively [16].

Employment of mothers was also found to influence the frequency of breastfeeding [5]. In Bahrain, 54% of employed mothers breastfed their infants on a schedule, compared with only 20% among nonemployed mothers [16]. This may be attributed to the positive association between the mother's employment status and educational level. In all Arab Gulf states, the majority of employed mothers are those with higher education [17].

Based on the National Child Health Surveys, the average frequency of breastfeeding in all six Gulf states was 5 times during the day and 3 times at night [6]. These figures are considered low when compared to the frequency of feeding in other developing countries [18]. Nevertheless, the studies in the Gulf did not show any association between the infant's age and feeding frequency. It is well known that as the infant grows older the frequency of breastfeeding decreases. This aspect is rarely given attention in studies related to breastfeeding in the Gulf.

A detailed study of the frequency of breastfeeding in Saudi Arabia was carried out by Madani et al. [19]. 24.7% of mothers breastfed every 2 h, and 11.5% every 3 h. Of mothers who remembered the duration of suckling, 12.1% mentioned that their children were suckled for <5 min, 52.5% for 5–10 min and the rest for >10 min. About 96% of mothers breastfed their children at night, and 88.1% of these breastfed their children more than once at night. The frequency of suckling per day and per night were significantly related to the mothers' education. However, the change of breasts at each feeding was not significantly related to the mothers' education or age.

Duration of Breastfeeding

In general, the duration of breastfeeding has declined steeply in the Gulf region. This decrease was more pronounced during the period 1973–1980 than from 1981 to 1990. 1973–1980 brought rapid changes in socioeconomic conditions and lifestyle in the Gulf, which negatively affected the duration of breastfeeding. In Bahrain, for example, the duration of breastfeeding was 2 years in the 1960s, but declined to 11 months in the 1970s and to 8 months in the 1980s [16]. The World Health Organization (WHO) [20] reported that traditionally, new mothers learned to breastfeed from their mothers and role models in the community. However, urbanization and economic transition have taken many women out of their traditional social support systems, so that successful role models of breastfeeding are no longer readily available. In the Gulf, more and more mothers believe that bottle feeding is a symbol of Westernization and they therefore introduce it when the infant is quite young. This, of course, has shortened the duration of breastfeeding.

The age at which breastfeeding stops in three Arab Gulf states (Bahrain, Oman and the UAE) is given in table 2. The patterns were very similar. About a third of children were breastfed for 1–3 months, and a quarter for 4–6 months. Only 24% of children in Bahrain were breastfed for 12 months, compared to 22.7% in the UAE and 19.4% in Oman.

It is worth mentioning that the mean duration of breastfeeding varies from country to country, and from region to region in the same Gulf country. The National Child Health Surveys in the Arabian Gulf showed that the average age at which breastfeeding stopped ranged from 9.9 months in Kuwait to 15.9 months in Oman [6]. These figures are not consistent with many community surveys which were carried out in the Gulf during the past 5 years. For instance, in Kuwait it was shown that the mean duration of breastfeeding was 5.8 months [21]. In Bahrain,

Table 2. Age at which breastfeeding stops in three Arab Gulf states

Age months	Bahrain [16]		Oman [13]		UAE [22]	
	n	%	n	%	n	%
1–3	76	28.6	28	27.2	75	33.3
4–6	54	20.3	24	23.3	53	23.6
7–9	42	15.8	17	16.5	27	12.0
10–12	30	11.3	14	13.6	19	8.4
13+	64	24.0	20	19.4	51	22.7
Total	266	100.0	103	100.0	225	100.0

Musaiger [16] found that the mean duration of breastfeeding was 8.8 months, which is similar to that in Qater (9 months).

This variation between different studies may be due to the method of calculating the mean, because most of the studies did not include mothers who were still breastfeeding at the time of the survey in the calculation. Thus the duration of breastfeeding may be underestimated. Ideally, the 'life-table' technique should be employed. However, this requires a special computer program and training which are not available in the region. The ability of the mother to recall correctly will also influence the findings. This is a particular problem when the target mothers are those who have children who are more than 2 years of age.

Bottle Feeding Practices

Bottle feeding is introduced as early as the first day of an infant's life. In many cases, infants receive formula before the mother's milk. This is mainly due to the mispractice in maternity hospitals of separating the infant from the mother after birth and providing formula for the newborn. In Bahrain, 54% of mothers stated that formula was the first food given to their infant [16]. The recent National Nutrition Survey in the UAE showed that 42% of mothers gave their children the bottle in the first, 21% in the second and 9% in the third month [22].

It is widely believed that early introduction of bottle feeding is one of the main factors affecting the continuation of breastfeeding in the Gulf region. Unfortunately, such practices start in the maternity hospitals and this may complicate the problem, since the mother has confidence in the practices of nurses and physicians working in these hospitals. In Bahrain, 68% of mothers used infant formula following advice of health staff [23].

Early introduction of the bottle can be hazardous. First, it influences the infants suckling. The two maternal reflexes involved in lactation are milk production and the milk rejection reflex, and both are responsive to suckling [24]. Second, bottle feeding from the first days carries the potential risk of infection. Unhygienic preparation of infant formula and feeding bottles may be major causes of diarrheal diseases and are associated with malnutrition in infants in the Gulf [2]. Third, studies in the region showed that many mothers prepared the infant formula incorrectly. In Saudi Arabia, 74% of mothers attending health centers prepared the feeds incorrectly, and 60% of them prepared overconcentrated feeds [25]. In Kuwait, it was reported that about 45% of mothers measured the scoop of powdered milk carefully [26]. Overconcentration or overdilution of powdered milk may adversely affect the health and nutritional status of the infant if formula is used for a long time.

Breastfeeding and Fertility

Numerous studies have demonstrated the beneficial role of breastfeeding in birth spacing. Frequent and exclusive breastfeeding are important factors in delaying postpartum fertility [27]. However, breastfeeding does not substitute for other contraceptives, but complements them, providing protection in the early months postpartum, when its contraceptive effect is strongest [8].

Information on the association between breastfeeding and the duration of postpartum amenorrhea (absence of menstruation during the period immediately following parturition) in the Gulf population is scanty. In Qatar, Musaiger et al. [unpubl. data] found that the mean duration of postpartum amenorrhea was 5.1 months. This is considered very short when compared with data in many developing countries [28]. In general, the mean duration of postpartum amenorrhea in Qatar increased as the duration of breastfeeding increased. It lasted 3.6 months in mothers who breastfed for < 1 month, increasing to 6.9 months among those who breastfed for > 12 months. In addition, the mean duration of postpartum amenorrhea among mothers who breastfed their babies was double that of mothers who did not breastfeed (5.5 and 2.6 months, respectively).

Breastfeeding can be a good natural contraceptive method during the first 3–6 months after delivery, but the use of other contraceptives soon after birth may help to shorten the duration of postpartum amenorrhea [11]. The above study in Qatar showed that the mean duration of postpartum amenorrhea was 4.2 months among mothers who used contraceptives compared to 7.9 months among those who did not. In Saudi Arabia, Al-Sekait [29] found that the difference in the mean duration of breastfeeding between users and nonusers of contraceptives was statistically significant (7 and 12.6 months, respectively).

In Oman, a recent study indicated that mothers who breastfed their infants during the first week had a longer mean duration of postpartum amenorrhea (6.4 months) than those who used bottle feeding (5.6 months) and mixed feeding (4.1 months). In addition, mothers who gave colostrum to their infants had a slightly higher mean duration of postpartum amenorrhea than mothers who did not (6.3 and 5.9 months, respectively) [9].

Table 3. Reasons for breastfeeding cessation in the Arabian Gulf countries

Reason		Bahrain [16] %	Kuwait [21] %	Oman [13] %	Qatar [49] %	Saudi Arabia [65] %	UAE [22] %
New pregnancy		22.7	14.5	39.8	7.0	40.1	16.4
Insufficient milk		20.5	30.3	11.7	35.0	12.2	26.7
Infant refusal		15.5	10.3	16.5	–	12.1	20.9
Infant reached weaning age		24.5	12.1	–	27.0	16.0	11.1
Mother's or child's illness		14.1	11.9	5.8	10.0	6.6	19.1
Other		2.7	20.9	26.2	21.0	13.0	5.8
Total	%	100.0	100.0	100.0	100.0	100.0	100.0
	n	266	2,994	103	1,121	611	255

Reasons for Stopping Breastfeeding

Factors determining the discontinuation of breastfeeding in the Gulf have been well documented. Studies conducted at the begining of the 1980s showed that a new pregnancy was the main reason for early cessation of breastfeeding. Nowadays, this factor ranks second in most Gulf States (table 3). This is probably due to the increase in the awareness of mothers as well as to the increased practice of family planning.

The belief that a pregnant mother's breast milk is 'bad' for the baby is widely accepted in the Gulf and a mother therefore stops breastfeeding as soon as she knows she is pregnant [30]. Supposed changes in breast milk volume or composition associated with a new pregnancy have not been confirmed by factual observations. It is unlikely that a nursing mother will become pregnant before her infant has begun to be weaned. When the infant is receiving supplementary food, and thus the frequency and intensity of suckling has decreased, the mother becomes pregnant. This usually happens when the infant reaches the age of 6 months, and another 2–3 months will usually elapse before the mother realizes that she is pregnant. At this time, she faces the decision whether or not to continue breastfeeding. However, by now the infant has benefited from most of the advantages of breast milk, and family foods can be easily and safely introduced [14].

'Lack of milk secretion' or 'insufficient milk' was the second reason why mothers stopped breastfeeding. The percentage of mothers who mentioned this reason was highest in Qatar (35%) and lowest in Oman (11.7%). Most mothers assume that they have insufficient milk because their babies are always crying or appear hungry, and they have insufficient knowledge as to why this has happened and how to deal with it. Regrettably, many mothers respond in a way that aggravates the problem, by giving the bottle to their babies [31]. If a mother truly

Table 4. Rate at which breastfeeding stops in Arabian Gulf countries

	Bahrain [35]		Kuwait [21]		Oman [13]		Qatar [unpubl. data]		UAE [22]	
	n	%	n	%	n	%	n	%	n	%
Gradually	138	55.5	1,378	57.0	55	41.5	150	66.0	107	55.4
Abruptly	172	44.5	1,058	43.0	39	58.5	76	34.0	86	44.6
Total	310	100.0	1,436	100.0	94	100.0	226	100.0	193	100.0

believes that she can provide milk for her infant, she will encounter some difficulties with milk let down [14]. However, many mothers need guidance and advice on the management of breastfeeding in this condition, and these are rarely provided in maternity hospitals.

'Infant reached weaning age' is another important reason mentioned by mothers in the Gulf. Most mothers believe that the child should stop breastfeeding by the end of the second year, as recommended by the Holy Koran. The majority of these mothers fail to continue breastfeeding for 2 years.

Traditional methods to stop breastfeeding are widely practiced in the Gulf region, especially in rural and bedouin communities. In Saudi Arabia, it was reported that the most common method is to put myrrh or black pepper on the mother's breasts. Among bedouins, another method still commonly used is to prick the infant's nose with a pin every time it approaches the breast [3]. Two studies in Saudi Arabia [32, 33] showed that putting bitter substances on the breast is a common method to stop breastfeeding. In Bahrain, several traditional methods are used to stop breastfeeding; the most widely practiced are putting bitter substances or iodine on the breasts, and covering the nipples with plasters [34].

A relatively large percentage of mothers in the region stop breastfeeding abruptly. The proportion ranged from 34 to 58.5% (table 4). However, in the rural and bedouin areas the percentage was much higher. For instance, in Bahrain, it was found that 73% of mothers in rural areas used abrupt methods to stop breastfeeding, compared to 39% in urban areas [35]. In bedouin communities of Saudi Arabia, almost all the mothers stopped breastfeeding abruptly [3].

Breastfeeding and Gastroenteritis

The importance of breastfeeding in the prevention of gastroenteritis has been well documented [36]. Gastroenteritis is considered to be one of the most common diseases among infants and young children in the Gulf countries. The increase in the practice of bottle feeding is more likely to raise the incidence of gastroenteritis among these children. Illiteracy, ignorance and unhygienic preparation of infant feeds are the main reasons for the prevalence of gastroenteritis among infants in these countries [2]. In Bahrain, Musaiger and Alshehabi [37] found that only 14.3% of mothers boiled the bottle after each feed, while 31.5% boiled it once a day, 32.1% twice a day and 10.1% did not boil the bottle at all. In Kuwait, Portoian-Shuhaiber and Al-Rashid [26] showed that 90% of mothers boiled the water for preparation of milk for their infants, but only 15% boiled water for drinking. Unhygienic preparation might have been done by foreign housemaids, since they are generally responsible for feeding the children, including those who have to be bottlefed. This may affect the health status of the children, because these housemaids have little or no knowledge of sterilization techniques [38].

A study in Kuwait among children aged 3–24 months indicated that the frequency and severity of gastroenteritis was greater among bottle-fed babies than breastfed babies. The percentage of bottle-fed babies who had one attack of gastroenteritis was almost double that of breastfed babies 3–5 and 6–8 months of age. The prevalence of gastroenteritis has increased among breastfed babies aged 9–11 and 12–24 months, but was still lower than among bottle-fed babies of the same age. This was attributed to the introduction of weaning foods [39].

El-Dosary et al. [40] surveyed children aged 1 month to 5 years who were suffering from acute gastroenteritis and admitted to two hospitals in Kuwait. They found that the incidence of gastroenteritis during the first year was higher (63.1%) among bottle-fed infants than infants fed breast milk or mixed feeds (36.9%). Moreover, 75.5% of children who had gastroenteritis were breastfed for <5 months. Children with the highest percentage of recurrent attacks of gastroenteritis (29.5%) were found among those with the shortest duration of breastfeeding (<2 months) compared to 15.4% among infants breastfed for >6 months.

Weaning Practices

The process of weaning varies from community to community and is often affected by food availability and economic and cultural factors. Weaning can be a dangerous period for babies. Inappropriate choice and unhygienic preparation of weaning foods may lead to infection and malnutrition. Indicators showed that many children in the Gulf failed to grow after 6 months of age [9, 41, 42]. In Bahrain, Zaghloul and Dodani [41] found that mean weight and height of children during the first 6 months of life were close to international standards, but were followed by faltering until about the end of the second year. This faltering is probably the result of adverse environmental factors and unsound weaning habits.

Most mothers introduced weaning foods quite early. Some mothers offered 'ghee' (melted clarified butter), dates and honey from the first days of life. Glucose and herbal water are also presented during the first week in the belief that they will relieve colic pain and treat constipation. Recent studies indicate that about 41 and 11% of mothers in Oman and Bahrain, respectively, gave glucose water to infants during the first 3 days of life [9, 16]. It was demonstrated that giving supplementary fluids such as water or glucose water in addition to breast milk were associated with a significant increase in the risk of diarrheal diseases [43].

The most common weaning foods used in the Gulf States are commercial preparations. These foods are usually introduced between the infant's second and third month. The National Nutrition Survey in the UAE showed that 21% of mothers introduced supplementary foods by the first, 68% by the second and 11% by the third month [44]. A study in Qatar reported that 50% of mothers introduced commercial weaning foods as the first foods given to their infants, followed by fruit juice (18.1%) [Musaiger et al., unpubl. data]. In Bahrain, it was shown that 74% of mothers gave commercial baby foods (including powdered milk) to their infants, and urban mothers were more likely to use these foods (77%) than rural mothers (58%) [35]. Infants who begin to eat weaning foods before they are 4 months old usually take less breast milk, because their small stomachs are easily filled, and therefore they may not grow well [45].

El-Sayed [46] provided information on the different supplementary foods given to children attending health centers in Riyadh (the capital), Saudi Arabia. About 53% of infants received fruits and vegetables at age 3–5 months. Fruits were given as juices or mixed with cereals and milk. Vegetables were cooked, mashed and mixed with their soups. Most of the infants were given rice pudding (*mehallabia*) and yoghurt. Of the other infants, 41% received the yolk of boiled eggs either alone or mixed with canned baby cereals. A minority of mothers (7%) gave legumes or minced meat (4%) at this period. By the age of 6–8 months the feeding pattern changed as infants were given most food items, particularly starchy foods such as rice, bread and potatoes, as well as meat, poultry and legumes.

Detailed information on weaning foods and age at their introduction in Oman has been reported [9]. Infant formula, glucose water, honey and commercial weaning foods were the main items introduced during the first 3 months of life. By 4 months all other foods were introduced and the most common were cereals, fruits, vegetables, chicken, meat, fish and dates. Interestingly, 6 and 12% of mothers introduced ghee (*samen*) at 1–3 and 4–6 months, respectively. They believe that ghee helps to strengthen the baby and prepares it for the taste of ghee in the future [47]. In Saudi Arabia, Sebai [3] found that 65% of settled, 94% of semisettled and 92% of the nomad mothers gave their children ghee at birth for the first 3 days of life. They believed that ghee lubricates the intestine, cleans it and gives the child nourishment. When ghee was not available, drops of caster oil were given instead.

The popularity of commercial weaning foods may be due to several factors: convenience, a family's high purchasing power, wide availability on the market, modern progress and development of these products and the image they are given through advertisements [48]. If the mother uses commercial weaning foods properly and at the right age, some of them may be good, but such foods should be used in addition to other home-based weaning foods. However, in the Gulf many mothers do not know how to prepare commercial weaning foods correctly and hygienically. Moreover, many mothers, especially in rural areas, did not use boiled water to prepare weaning foods, which increase the risk of diarrheal diseases [2].

Factors Influencing Breastfeeding in the Gulf

Breastfeeding pattern is affected by various biological, social, cultural and economic factors. In the Gulf region most of the studies have focused on the impact of social and economic factors on breastfeeding. There is a lack of information on the health and cultural factors associated with infant feeding practices. The most important factors determining breastfeeding in the Gulf based on published literature are summarized below.

Age of Mother
The mother's age may be associated with two factors: education and experience. Studies in the Gulf have indicated that as the mother's age increases, the educational level decreases [49–51]. Young mothers may have no experience in breastfeeding management and therefore may easily lose confidence and decide to feed the baby artificially [52]. The duration of breastfeeding in Saudi Arabia was significantly associated with the mother's age. The duration was 9.4, 10.6, 13.2 and 15.9 months for mothers of age groups 15–24, 25–35, 36–44 and 45+ years, respectively [29]. A similar trend was observed in Kuwait, with the mean duration of breastfeeding highest among mothers aged 35 years and over (7.9 months), compared to 4.4, 5.0 and 6.6 months for mothers <20 years, 20–30 and 30–35 years, respectively [21].

Education of Mother
The mother's educational level has a potent effect on infant feeding practices in the Gulf. Most of the mothers in this region are either illiterate or have a low educational level (primary school), and this may unfavorably influence the food habits of young children and other family members. However, education is not always connected with good food habits. This is particularly noticable in infant feeding practices, where the mother's educational level was negatively correlated with the duration of breastfeeding and age of introducing the bottle.

Table 5. Mean and median duration (months) of breastfeeding by educational level of mothers in the Arab Gulf countries

Country	n		Educational level of mother				Reference
			illiterate	primary and intermediate	secondary	university	
Bahrain	262	mean	9.2	8.2	6.6	6.5	16
		median	6.0	6.0	5.0	4.5	
Kuwait	2,994	mean	9.5	5.6	4.1	4.4	21
		median	–	–	–	–	
Oman	217	mean	10.8	9.4	5.5	6.1	9
		median	12.0	8.0	5.0	5.0	
Qatar	362	mean	13.6	9.7	8.6	7.6	unpublished
		median	12.0	9.0	7.0	6.0	
Saudi Arabia	2,010	mean	13.6	10.3	8.6	5.3	29
		median	–	–	–	–	
UAE	225	mean	8.8	7.3	7.2	5.8	22
		median	6.0	4.5	6.0	4.0	

In Saudi Arabia it was found that the duration of breastfeeding was 13.6 months among illiterate mothers and decreased sharply to 5.3 months among mothers with a university education. In Kuwait, the duration decreased from 9.5 to 4.3 months among illiterate and highly educated mothers, respectively [21]. A similar pattern was noted in other Gulf countries (table 5).

Geographical Distribution

Breastfeeding patterns differ widely by geographical area. In general, the mean duration of breastfeeding is higher in rural or bedouin communities than in urban or settled communities. Serenius et al. [53] studied breastfeeding in four areas in Saudi Arabia: privileged urban, average urban, less privileged urban and rural areas. The data were analyzed using a life-table technique. The median duration of breastfeeding in these four groups was 2.1, 7.5, 10.8 and 17.8 months, respectively. By 3 months of age, 90% or rural children were breastfed. However, at this age, 42% of privileged, 68% of less privileged and 74% of average urban children were breastfed. When the children reached 12 months of age, 67% of them in rural areas were still breastfed, whereas the proportion had declined remarkably in the other groups. The introduction of solid foods started earlier for urban than for rural children. At 6 months, only 7% of rural children were on solid foods compared with 90% of privileged urban children. At 1 month, 52% of privileged and 42% of less privileged urban children received bottle feeding, while the percentages were 16 and 17% in rural and average urban children, respectively.

In general, geographical differences in breastfeeding patterns can be best observed in Saudi Arabia, Oman and the UAE, with large land masses and varied topographies comprising coastal, mountain and desert areas.

Womens' Employment

Although women's enrollment in work has increased annually in the Gulf, they still represent a small proportion of the work force. The percentage of employed women (those earning money from part of their work), especially nationals, is low in all Gulf countries, ranging from 5 to 10% [54]. Employed women can be classified into two groups: those who are covered by maternity leave and those who are not. Most women working in formal sectors as well as in some private sectors in the Gulf are entitled to maternity leave ranging from 45 to 70 days. Actually, the majority of lactating women extend this leave by taking their annual leave of 20–40 days. Therefore, the woman has about 2–4 months to breastfeed her child at home. In some Gulf countries women are entitled to an hour a day during working hours for breastfeeding.

Employment is rarely mentioned as a reason for stopping breastfeeding in the Gulf, but is often mentioned as a reason for early introduction of the bottle and weaning foods. Overall, the available data indicate that the duration of breastfeeding is highly associated with womens' employment. In Saudi Arabia and Kuwait, the duration of breastfeeding was significantly less among employed than unemployed women (table 6). However, in Bahrain, the situation is reversed; the mean duration was higher among employed (10.3 months), than unemployed (7.9 months) women. There was no clear explanation for this difference, but it may be due to an increasing awareness among employed women about the importance of continuing to breastfeed. The keenness of the employed woman to bond with her baby for as long as possible may contribute in part to the increased duration of breastfeeding, because many women feel guilty about leaving the child at home with foreign housemaids, particularly when they find that the child has become more linked to the housemaid than the mother.

Influence of Housemaids

The presence of maids in Gulf homes has become a phenomenon rather than a real need. Gulf families have become more and more dependent on the services of housemaids, who are often poorly qualified. High incomes and the gradual employment of women, especially from nuclear families, called for the service of housemaids. Most of these housemaids have had little education, and few of them can understand or speak Arabic. They come from totally different cultures such as the Indian subcontinent and the Far East. They are responsible for all levels of

Table 6. Mean duration of breastfeeding by employment status of mother in the Arab Gulf states

Country	Employment status	Sample size	Duration of breastfeeding months (mean ± SD)	Reference
Bahrain	employed	51	10.3 ± 7.5	16
	not employed	215	7.9 ± 6.2	
	total	266	8.4 ± 6.5	
Kuwait[a]	management	–	3.8 ± 1.8	21
	technical	–	4.4 ± 2.3	
	not employed	–	7.0 ± 2.8	
	total	2,994	5.8 ± 3.4	
Oman	employed	17	5.6 ± 4.0	9
	not employed	200	10.0 ± 6.6	
	total	217	9.7 ± 6.5	
Qatar	employed	79	10.9 ± 8.1	unpublished
	not employed	147	12.1 ± 8.7	
	total	226	11.7 ± 8.5	
Saudi Arabia	employed	266	6.0 ± 2.5	29
	not employed	1,744	12.8 ± 2.3	
	total	2,010	11.4 ± 3.1	
UAE	employed	21	6.7 ± 6.0	22
	not employed	204	8.3 ± 7.1	
	total	225	8.1 ± 7.0	

[a]Sample size of groups was not reported.

home management, including feeding the infants and young children. It is therefore widely accepted that these housemaids have a big influence on the food habits of children in the Gulf [55].

Amine and Al-Awadi [56] examined the influence of maids on infant feeding practices in Kuwait. It was found that in the presence of maids, 55% of the infants were breastfed from birth, 18% were bottle fed and 27% received mixed feedings. In the absence of maids, the percentage of breastfeeding increased to 65%, and that of bottle and mixed feeding decreased to 13 and 22%, respectively ($p < 0.01$). The duration of breastfeeding was 7.3 months in the absence of maids, and declined to 5.4 months in their presence ($p < 0.01$). The average age at introduction of bottle feeding decreased significantly in the presence of maids. However, it is difficult to conclude that this change in infant feeding practices was due to the presence of maids. It is more likely that homes with maids include employed and highly educated mothers, as well as having a high family income. These are the most confounding factors affecting infant feeding practices, rather than the maid alone.

The Role of Health Workers
Health workers who are in direct contact with mothers play an important role in promoting breastfeeding and sound weaning habits. Their knowledge and attitudes are reflected by the mother attending maternal and child health (MCH) clinics or maternity hospitals. Unfortunately, a large proportion of these workers lack sound information on infant feeding practices. A study of the knowledge and attitudes of health workers in five health centers in Oman showed that 36% of them recommended the schedule method of breastfeeding. About 6% believed that weaning should start at the second month, 30% at the third month and the rest at the fourth month or later. Eleven percent of the health workers recommended that breastfeeding should be discontinued when the infant has diarrhea [13].

Private doctors in the Gulf are also a source of worry. Many private clinics are becoming good places for promoting infant formulas and commercial weaning foods. The food companies offer samples, leaflets, toys and medical equipment free of charge to the private clinics in order to promote their products. The surrounding environment in these clinics is full of advertisements on various types (and sometime only one type) of baby foods. This will certainly influence the mother who attends such clinics.

Sex of the Child
Several studies in Arab regions have demonstrated that boys are breastfed for longer periods than girls [5]. Gender discrimination is considered to be the cause of the poorer health and nutritional status of girls in many developing countries [57]. In the Gulf, few studies reported infant feeding practices by sex of the child. Indications show that the mean duration of breastfeeding was higher among boys in Oman and Bahrain, but the difference was not significant [9, 16]. In Kuwait, it was found that the infant's sex did not have a significant association with breastfeeding patterns, weaning practices and the age at which the bottle is introduced [21].

In Saudi Arabia, Anokute [58] reported that the desire to have another pregnancy was more often given as a reason for stopping breastfeeding by mothers who had delivered female rather than male babies (66.4 vs. 33.4%, respectively). The investigator believed that the male preference may play an important role in the decision to stop breastfeeding. However, further investigations are required to determine whether or not there is sex discrimination in breastfeeding practices in the Gulf.

Other Factors

Many other factors could affect breastfeeding patterns in the region. The type and place of delivery were found to be significantly associated with breastfeeding in Saudi Arabia [29]. Mothers with normal delivery had a higher mean duration of breastfeeding (11.7 months) than mothers with an abnormal delivery (9.8 months). The mean duration of breastfeeding was higher among mothers who delivered at home (12.6 months) than those who delivered in hospitals (9.5 months). In Oman, Musaiger [9] reported that the mean duration of postpartum amenorrhea was longer among mothers who delivered their babies at home compared to those who delivered in hospital (7.2 and 6.2 months, respectively). This is probably because mothers who delivered at home were less likely to be exposed to the promotion of infant formula.

Although, in the Gulf, income is a less sensitive indicator of socioeconomic status than education [59], some investigations have shown that family income correlates negatively with the duration of breastfeeding and age for introducing bottle feeding [21, 29, 60]. But, income is usually linked with education, and it is the parents' education rather than their income that affects infant feeding practices.

The birth weight of infants was found to be associated with their feeding patterns. In Bahrain, the duration of breastfeeding was found to be less for infants with low birth weight (<2.5 kg) than for normal infants (7.6 months for low-birth-weight infants compared to 8.9 and 10.4 months for infants with birth weights of 3 and 4 kg, respectively). In addition, low-birth-weight infants were less likely to be breastfed immediately after birth than normal infants [unpubl. data].

Marketing of Baby Foods in the Gulf

The Gulf region has been subjected to massive promotion and marketing campaigns by baby food companies. The availability of a wide variety of infant formulas and other baby foods in the Gulf and high purchasing power play an important role in the use of these foods at a very early stage in the infant's life. In Bahrain, there were 203 types of baby foods available on the market, of which 17 were infant formulas and 37 were cereal-based products [61]. In Saudi Arabia, the number of infant formulas available on the market was much higher than in Bahrain. For example, Al-Frayh [62] reviewed more than 36 infant formulas in the city of Riyadh alone. A few years ago, most of these formulas did not follow the nutritional standards recommended by the Food and Agricultural Organization (FAO)/WHO. In 1978, a study carried out on infant foods available in Saudi Arabia showed that 50% of the formulas had a higher protein content than was recommended. Sodium figures were generally not stated on the label. Of 31 formulas on sale, only 14 met the FAO/WHO Codex Standard [63]. The situation has improved recently, since most of the Gulf countries are regulating the marketing of infant formulas.

Musaiger and Lankarani [61] studied the properties of infant formulas and weaning foods available in Bahrain. During the survey there were 17 formulas for use in infant feeding, produced by eight manufacturers from seven countries. All of these formulas (except one) were based on cow's milk which was modified and supplemented with vitamins and minerals. The milk fat is partially replaced with vegetable oils in the other formula. The average content of minerals and vitamins per 100 kcal based on labelling information showed that all these formulas were within the range recommended by the Codex Alimentarius Commission [64]. However, the findings indicated that there was wide variation in mineral and vitamin content among the different formulas. Sodium, for example, ranged from 22 mg/100 kcal in one product to 61 mg/100 kcal in another, while that of niacin ranged from 269 to 1,754 μg/100 kcal.

The information declared on the label of infant formulas and commercial weaning foods available in Bahrain and Oman has been studied [13, 61]. Production and expiry dates are generally marked on most baby foods. Arabic descriptions were available for all infant formulas and baby cereals, but in Bahrain, 50% of bottled weaning foods had no information in Arabic on their labels while in Oman, most of bottled weaning foods had no Arabic information. The Codex Alimentarius Commission [64] recommended that the language used on food labels should be a language acceptable to the country in which the food is intended for sale. For the Gulf, this language should be Arabic, in addition to another acceptable language (English). Most of the Arabic information was typically translated from English. This kind of translation is misleading because it does not give the proper meaning in Arabic. Clear and understandable Arabic information should be declared on the labels. This can be achieved by joint efforts between professional staff and someone who understands Arabic very well [61].

Storage instructions are also important, because consumers should know how to store baby foods safely, especially after opening. Although most households in the Gulf have refrigerators, many mothers do not know when and where to store the milk and other foods given to the child. For example, in Saudi Arabia it was found that 97% of mothers kept leftover baby milk at room temperature which may lead to its spoilage [65]. The climate of the Gulf (high humidity and high temperature) provides unsuitable conditions for storing foods outside the refrigerator. Therefore, information concerning the best place or temperature for storing baby foods, especially after opening, is essential in the region. In Bahrain [61], all infant formulas had such information, but 100% of cereal mixed with vegetables, 92% of cereal mixed with fruits and 37% of cereal products did not have this information.

Firebrace [38] reported that it was standard practice for all mothers in the Gulf when discharged from maternity hospital to leave with a free formula sample. This practice has meanwhile stopped in most hospitals, but company representatives still visit hospitals and private clinics leaving samples with the medical or administrative staff. In the UAE, Osman [66] found that about 5–19% of infants were given a formula immediately after delivery. Autret and Miladi [67] noticed that infant formulas were distributed free to mothers who attended health centers in the Gulf countries. This distribution is now regulated and free samples are restricted to those mothers who cannot breastfeed their children or to poor mothers who cannot afford infant formulas. Nevertheless, the practice of giving free samples of infant formulas encourages mothers to use them and hence discontinue breastfeeding. This was demonstrated by Bergerin et al. [68], who found that those mothers who received infant formula samples were less likely to continue breastfeeding during the first month of the infant's life and would probably introduce solid foods by the second month. This trend becomes more significant among less educated mothers, primiparas and mothers who have been ill postpartum.

Advertisements also partially influence mothers to use baby foods. Although infant formula advertisements were banned from the mass media, in all Gulf countries other advertising techniques are pursued, for example the distribution of leaflets, posters, calendars, and free samples or window displays of infant formulas in shops. Many private clinics are used for promoting infant formulas. Free samples of certain baby foods are still being given by private doctors. The waiting rooms in these clinics are full of commercial infant food advertisements which attract the attention of mothers. In Saudi Arabia, Haque [69] found that mothers used six varieties of infant formula for their babies before they reach the fourth month of age.

The cost of bottle feeding and canned weaning foods is not taken into consideration by most families in the Gulf. This is particularly true with the middle- and high-income groups, to which the majority of families in the region belong. Musaiger and Lankarani [61] have attempted to study the cost of 100 g protein and 1,000 kcal of infant formula and canned baby cereals available in Bahrain. They found that the price of 100 g protein of infant formula ranged from US$2.60 to 7.60, while that of 1,000 kcal ranged from US$1.00 to 2.10. For canned baby cereals, the price of 100 g protein varies from US$4.05 to 9.90, and for 1,000 kcal from US$1.03 to 2.70. They concluded that the relative prices for canned baby cereals were much higher than those for fresh foods. The price of 100 g protein is 2- to 7-fold more expensive than for local foods such as meat, chicken, wheat flour and rice. The price of 1,000 kcal for wheat flour is one thirteenth the price in canned cereal foods for children. Interestingly, the cost of cereal protein in baby foods was higher than that in infant formulas, although the latter are higher in quality.

In Oman, the cost of bottle feeding and weaning foods was higher than that reported in Bahrain. 100 g protein of infant formula ranged from US$3.80 to 9.80 and 1,000 kcal ranged from US$1.50 to 2.70. The price range of canned weaning foods for 100 g protein was US$4.60–13.00, while that of 1,000 kcal was US$1.40–2.90 [13].

Activities to Encourage and Support Breastfeeding in the Gulf

Several activities and programs are carried out in the Gulf to protect and support breastfeeding practices. These are summarized below.

Maternity Protection

As stated earlier, employment of the mother constitutes an important factor responsible for the early decline of breastfeeding. The Gulf States allow female workers to have maternity leave with full pay for not less than 45 days, which should not be deducted from their annual leave. In Kuwait and Saudi Arabia, maternity leave is about 2 months with full pay. Most of the Gulf States allow nursing mothers to extend their maternity leave for up to 1 year without pay. In Kuwait, up to 4 years without pay are allowed for Kuwaiti mothers only. The economic situation in the Gulf is generally good, hence many working mothers can stay at home without facing shortages in the household budget [70].

In Bahrain, a nursing mother is allowed 1 h each day to breastfeed her child at home. The timing is subject to the circumstances of the mother and the employer. In Kuwait, the nursing break during the working day is optional, and depends on an agreement between the employer and the mother (table 7).

Educational Activities

Serious efforts have been made by the Secretariat of Health for the Arab Gulf states to improve health and nutrition education, as well as public awareness in the region. As a result, a health program called Salametek is broadcast through television and radio networks in the Gulf. It comprises 52 films of 25 min each for television, and 52 talks of 15 min each for radio, supported by 260 messages of 2–3 min daily on television and radio. A balanced diet for pregnant and lactating mothers, the advantages of breastfeeding and sound weaning habits were given priority in this educational program.

Posters, leaflets and booklets directed at pregnant and nursing mothers have been developed by the Gulf Ministries of Health, and distributed to hospitals, health centers and health practitioners in the region. The last 10 years have witnessed an increase in the promotion of breastfeeding campaigns in the mass media. There are already some activities being carried out in health and MCH

Table 7. Provisions to support mothers in the work force in the Arab Gulf countries [72]

Country	Maternity leave policy	Salary during leave	Provisions for nurseries/crèches	Nursing breaks
Bahrain	45 days in addition to annual leave	100% paid by employer	none	1 h per day; time counted as time worked; mothers are expected to commute home, but Ministry of Health states that most mothers live close to where they work
Kuwait	30 days before, 40 days after birth; up to 100 days extension allowed	100%; right to annual vacation is lost; paid by employer	none	1 h per day (optional)
Oman	45 days	100%	none	–
Saudi Arabia	4 weeks before, 6 weeks after birth; 6 months extension for complications	50% of salary if <3 years of service and 100% if ≥3 years	required in enterprises employing >50 women	1 h per day plus normal breaks; time counted as time worked
UAE	9 weeks; at least 12 months uninterrupted work with same employer to qualify; 100 days unpaid extension, consecutive or not, for complications; up to 100 days sick leave may be used for illness resulting from pregnancy or childbirth	100% for women with 1 year service; 50% for others	none	0.5 h twice a day for 18 months; time counted as time worked

Musaiger

184

centers in the form of nutrition demonstrations, exhibitions, home visiting, promotion of breastfeeding and advice on sound weaning practices.

Marketing Activities

At the beginning of 1982, the Gulf states decided to abandon all infant formula advertising on television. In some states, like Kuwait and Saudi Arabia, the ban extended to radio, newspapers and magazines. The Secretariat of Health for the Arab Gulf states initiated a committee to study the possibility of adopting the 'Code of Marketing of Breast-Milk Substitutes' suggested by WHO/UNICEF. The committee redrafted and modified the code. Each state has to adopt this modified code according to its social and political circumstances. It is likely that in the near future most of the Gulf states will respond to the code through legislation [70].

Urgent action was taken in some states to regulate the distribution of infant formulas in hospitals and health centers. All samples of infant formulas are now banned from free distribution to the mothers or health personnel in maternity hospitals and health centers. In 1983, the Saudi Arabian Standards Organization [71] released some regulations related to breast milk substitutes:

(1) Government authorities shall exert all possible efforts and cooperate in the implementation of the informational and educational materials recommended by the 'International Code of Marketing of Breast-Milk Substitutes', regarding the process and publicizing of breastfeeding.

(2) Government authorities shall encourage breastfeeding and develop an information program emphasizing its advantages, benefits and superiority.

(3) Government authorities shall prohibit all commercial promotion for breast-milk substitutes and give them no financial support.

(4) Manufacturers and distributors shall be requested to declare the following information in a clear and legible form on each container of breast milk substitute.

    (a) A statement indicating the superiority of breastfeeding, to be preceded by the phrase: 'Important Notice'.

    (b) Instructions for appropriate preparation, and a warning against the health hazards of improper use.

    (c) The label or container shall not bear any pictures of infants or other pictures or text which may idealize the use of infant formula.

    (d) The label shall state the ingredients used, the chemical composition, storage conditions required, the batch number, date of production and expiry date, if the latter is not > 18 months from the date of production.

Support of Appropriate Weaning Practices

Activities related to sound weaning practices are becoming a top priority in the health programs of the Gulf states. In Bahrain and Oman, for example, the decision was taken to standardize the information given to mothers regarding the appropriate age at which infants who are exclusively breastfed should be given additional foods. It was concluded that the available evidence and experience indicated that among healthy communities, exclusive breastfeeding is sufficient for the adequate growth of most infants for 4–6 months. Therefore, a guideline for the complementary foods given to infants during the first year of life was produced and is distributed to both health staff and the public.

Workshop and Seminar Activities

The Gulf States have begun to organize workshops, scientific meetings and symposia conducted at national or regional levels. These activities have been directed towards raising the awareness of health personnel about the importance of breastfeeding and the support measures needed to facilitate it. Training workshops, especially for nurses and MCH staff, are held annually to increase their skills in the management of breastfeeding. Some countries, like Oman and Bahrain, have prepared a training manual to promote and support breastfeeding. The manual is in two languages, Arabic and English, because many of the health staff are foreigners.

Research Activities

Up to 1979, research activities concerning infant feeding were very limited in the Gulf. In 1980, a joint FAO/UNICEF mission visited the Gulf states to make a first assessment of the nutrition situation, with particular reference to infants and young children. The recommendations of the mission emphasized the need for surveys of the infant feeding practices in the area. As a result, studies on infant feeding habits and the factors influencing them have been conducted in all Gulf states. The best examples are the Child Health Surveys which were carried out in all six Gulf states, and sponsored by many agencies, including the Gulf Health Council. These surveys covered demographic characteristics of mothers and children, family planning, infant and child mortality, diarrheal morbidity, accidents and disability, immunization, maternal health, and breastfeeding and weaning habits.

Conclusion

In general, the duration of breastfeeding has declined in most communities in the Arab Gulf region, and mixed feeding (breast and bottle) is the norm. However, the main problem in this region is probably the early introduction of several foods such as water, glucose water, herbal water, ghee and infant formula very early in the child's life. This increases the risk of infection as well as adversely affecting the frequency of suckling. The results are a short duration of breastfeeding and possible malnutrition.

Several programs are currently underway to promote breastfeeding in this region. Nevertheless, they have shown little effect on breastfeeding patterns, which may be a consequence of several factors: (1) lack of coordination and cooperation among various bodies involved in supporting breastfeeding; (2) lack of health regulations and legislation regarding infant formulas and other baby foods; (3) difficulty in controlling the private hospitals and clinics, which play an important role in encouraging bottle feeding; (4) insufficient knowledge among health personnel concerning the management of breastfeeding; (5) a severe shortage of behavioral and social studies related to breastfeeding, and (6) insufficient training of MCH nurses and other health workers in nutrition in general, and in sound infant feeding practices in particular.

To promote breastfeeding in the Gulf communities, a global policy should be initiated. It is worth mentioning that all Gulf countries recently established a so-called "National Child Health Plan" to promote children's health. Sound infant feeding is one of the main items in this plan, but the question arises, how practical and easy will it be to implement this plan, especially with the lack of research and studies on factors associated with breastfeeding. Breastfeeding cannot be promoted in this region without strict regulations regarding the marketing of infant formulas and other baby foods, adequate and effective training of health workers, well-designed mass media campaigns, introducing sufficient information on sound infant feeding practices in school and university curricula, and finally convincing policy makers about the benefits of breastfeeding to the health of child and mother and the role of breastfeeding in reducing the cost of health services.

References

1   Musaiger AO: Nutritional status of mothers and children in the Arab Gulf countries. Health Promot Int 1990;5:259–268.
2   Musaiger AO: The state of food and nutrition in the Arabian Gulf countries. World Rev Nutr Diet 1987;54:105–173.
3   Sebai ZA: The Health of the Family in a Changing Arabia, ed 4. Jedda, Tohama, 1984.
4   Elias JBT: A survey of place of delivery, modes of milk feeding and immunization in a primary health care centre in Saudi Arabia. Saudi Med J 1985;6:169–176.
5   Harfouche J, Musaiger AO: Breastfeeding Patterns: A Review of Studies in the Eastern Mediterranean Region, ed 2. Alexandria, World Health Organization, Regional Office for the Eastern Mediterranean Region, 1992.
6   Gulf Council of Health Ministries: The Gulf Child Health Survey: Basic Indicators. Riyadh, GCHM, 1990.
7   WHO/UNICEF: Breastfeeding in the 1990s. Geneva, World Health Organization, 1990.
8   Berg A, Brems S: A Case for Promoting Breastfeeding in Projects to Limit Fertility. Washington, World Bank, 1989.

9   Musaiger AO: Health and Nutritional Status of Omani Families. Muscat, UNICEF, 1992.
10   Almorth S, Bidinger PD: No need for water supplementation for exclusively breast-fed infants under hot and arid conditions. Trans R Soc Trop Med Hyg 1990;84:602–604.
11   Agency for International Development: Breastfeeding for Child Survival Strategy. Washington, Agency for International Development, 1990.
12   VanLandingham M, Trussell J, Grummer-Strawn L: Contraceptive and health benefits of breastfeeding; A review of the recent evidence. Int Fam Plann Perspect 1991;17:131–136.
13   Musaiger AO: A Rapid Assessment of Weaning Habits in Oman. Muscat, UNICEF, 1988.
14   Akre J (ed): Infant Feeding: The Physiological Basis. Bull World Health Organ 1989;67(suppl).
15   WHO/UNICEF: Protecting, Promoting and Supporting Breastfeeding. Geneva, World Health Organization, 1989.
16   Musaiger AO: Social Factors Affecting Breastfeeding in Bahrain (in Arabic). Bahrain, Ministry of Health, 1991.
17   Thabit N: Women, Development and Social Changes. Kuwait, That Al-Salasel, 1983.
18   WHO: Contemporary Patterns of Breast-feeding. Geneva, World Health Organization, 1991.
19   Madani KA, Al-Nowaisser A, Khashoggi RH: Lactation and fertility among Saudi women. Jedda, Ministry of Health, 1990.
20   WHO: Breastfeeding: Why the Global Concern Now? Geneva, World Health Organization, 1990.
21   Amine EK, Al-Awadi F, Rabie M: Infant feeding pattern and weaning practices in Kuwait. J Roy Soc Health 1989;109:178–180.
22   Musaiger AO: The State of Food, Nutrition and Health in the UAE. Abu-Dhabi, Ministry of Health, 1992.
23   Musaiger AO: The extent of bottle feeding in Bahrain. UNU Food Nutr Bull 1983;5:20–22.
24   Minchin M: Breastfeeding Matters. North Sydney, Allen & Unwin, 1985.
25   O'Donovan DJ, Riggall HM, Harland PSE: Dangerous infant feeding preparation techniques: The choice of maternal methods and improper formula in Riyadh, Saudi Arabia. J Trop Pediatr 1985;31:94–96.
26   Portoian-Shuhaiber S, Al-Rashid A: Feeding practices and electrolyte disturbances among infants admitted with acute diarrhoea – A survey in Kuwait. J Trop Pediatr 1986;32:168–173.
27   WHO: Breastfeeding and Child Spacing. Geneva, World Health Organization, 1988.
28   Saadeh R, Benbouzid D: Breastfeeding: The Implications for Child Spacing and Fertility. Geneva, World Health Organization, 1990.
29   Al-Sekait MA: A study of the factors influencing breastfeeding patterns in Saudi Arabia. Saudi Med J 1988;9:596–601.
30   Musaiger AO: Child Nutrition in the Arab Gulf (in Arabic). Kuwait, Kuwait Society for the Development of Arab Childhood, 1990.
31   Graffy J: Breastfeeding: the GP's role. Practitioner (East Mediterr Edit) 1992;236:549–551.
32   Al-Othaimeen AI, Villanuera BP, Devol EB: The present trend in infant feeding practices in Saudi Arabia. UNU Food Nutr Bull 1988;9:62–68.
33   Al-benian AS, Shana SA: Childhood Needs in the Saudi Arabian Community, in the Arab League: Principal Needs for the Child in the Arab World (in Arabic). Tunisia, Arab League, 1980, pp 178–190.
34   Department of Social Planning and Research: Weaning Habits in Bahrain (in Arabic). Manama, Ministry of Health, 1985.
35   Musaiger AO: Food habits in Bahrain: Infants' feeding habits. J Trop Pediatr 1983;29:248–251.
36   Feachem RG, Koblinsky MA: Intervention for the control of diarrhoeal diseases among young children: Promotion of breast-feeding. Bull World Health Organ 1984;62:271–291.
37   Musaiger AO, Alshehabi KH: Infant feeding practices in Bahrain: A pilot study; in Musaiger AO (ed): Studies on Nutrition in Bahrain. Manama, Ministry of Health, 1984, pp 5–7.
38   Firebrace J: Infant Feeding in the Middle East. New York, International Baby Food Action Network, 1983.
39   Anonymous: Possible hazards of artificial feeding in Kuwait. Monthly Epidemiological Report No 478. Kuwait, Ministry of Health, 1978.

40 El-Dosary L, Nour S, El-Sherbini AF, et al.: Ecological factors influencing morbidity patterns of acute gastroenteritis in children under five years of age in Kuwait. Bull High Inst Public Health Alexandria 1982;12:49–65.

41 Zaghloul NE, Dodani T: A study of the growth pattern of Bahraini children. Bull High Inst Public Health Alexandria 1984;14:147–156.

42 Musaiger AO, Al-Sayyed J: Nutritional Status of Mothers and Children in Bahrain. Manama, Ministry of Health, 1991.

43 Hofvander Y (ed): Maternal and Young Child Nutrition. Paris, UNESCO, 1983.

44 Ministry of Health: Results of the National Nutrition Survey. Abu Dhabi, Ministry of Health, 1992.

45 WHO: Weaning: From Breast Milk to Family Food. Geneva, World Health Organization, 1988.

46 El-Sayed NA: Infant feeding and weaning practices in Riyadh, S. Arabia. Bull High Inst Public Health Alexandria 1985;15:179–191.

47 Musaiger AO: Food and Nutrition in the Southern Region of Oman. Muscat, UNICEF, 1990.

48 UNICEF: Good Weaning Practices. Muscat, UNICEF, 1990.

49 Ministry of Health: Qatar: Child Health Survey. Doha, Ministry of Health, 1991.

50 Ministry of Health: Saudi Arabia: Child Health Survey. Riyadh, Ministry of Health, 1991.

51 Ministry of Health: Bahrain: Child Health Survey. Manama, Ministry of Health, 1992.

52 Savage F: The need for action. Int J Gynecol Obstet 1990;31(suppl):11–15.

53 Serenius F, Swailem AR, Edresse AW, et al.: Patterns of breastfeeding and weaning in Saudi Arabia. Acta Paediatr Suppl 1988;346:121–129.

54 Al-jerdawy AA: Problems of Kuwaiti and Gulf Employed Women (in Arabic). Kuwait, That Al-Salasel Publishers, 1986.

55 Musaiger AO: The newly industrializing countries: Nutrition education in the face of rapid change, Experience in Bahrain; in Taylor TG, Jenkins NK (eds): Proceedings of the XII International Congress of Nutrition. London, Libbey, 1986, pp 917–920.

56 Amine EK, Al-Awadi F: Expatriate maids and food patterns in Kuwait. J R Soc Health 1990;110:138–140.

57 UNICEF: The Girl Child. New York, UNICEF, 1991.

58 Anokute CC: Infant feeding in King Khalid University Hospital. J R Soc Health 1988;108:199–200.

59 Musaiger AO: Social and economic indicators for assessing nutritional status; in Musaiger AO (ed): Proceeding of the National Seminar on Food and Nutrition Surveillance. Manama, Ministry of Health, 1990, pp 18–24.

60 Al-Bustan MH: Attitudes and practice of Kuwaiti women towards breastfeeding. Int Q Commun Health Educ 1987;7:135–148.

61 Musaiger AO, Lankarani S: A Study of Baby Foods Available in Bahrain. Manama, Ministry of Health, 1986.

62 A-Frayh AS: Infant formula available in Saudi Arabia. Saudi Med J 1986;7:218–226.

63 Lawson M: Infant feeding habits in Riyadh. Saudi Med J 1981;2:26–29.

64 Codex Alimentarius Commission: Codex Standard for Foods for Special Dietary Uses Including Foods for Infants and Children, ed 1. Joint FAO/WHO Food Standard Programme, Vol 9. Rome, Food and Agriculture Organization, 1982.

65 Sawaya WN, Tannous RI, Al-Othiameen AI, et al.: Breast-feeding practices in Saudi Arabia. UNU Food Nutr Bull 1987;9:69–72.

66 Osman AK: Nutrition Status Survey: UAE. Abu-Dhabi, UNICEF, 1981.

67 Autrit M, Miladi S: Analysis of Services for Children in the Countries of the Gulf; in UNICEF: Proceedings of the First Regional Workshop on Nutrition as Related to Children and Mother Health in the Gulf Countries. Abu Dhabi, UNICEF, 1980, pp 25–41.

68 Bergerin Y, Dougherty C, Kramer MS: Do infant formula samples shorten the duration of breastfeeding? Lancet 1983;1:1148–1151.

69 Haque KN: Feeding pattern of children under two years of age in Riyadh, S. Arabia. Ann Trop Paediatr 1983;3:129–132.

70 Musaiger AO: Encouragement and support of breastfeeding in the Arab Gulf countries; in Jelliffe

DB, Jelliffe EFP (eds): Programmes to Promote Breastfeeding. Oxford, Oxford University Press, 1988, pp 145–149.

71 Saudi Arabian Standards Organization: Requirements for Breast-Milk Substitutes. Riyadh, Saudi Arabian Standards Organization, 1983.

72 Clearinghouse on Infant Feeding and Maternal Nutrition: Legislation and Policies to Support Maternal and Clinical Nutrition. Report No 9. Washington, American Public Health Association, 1989.

Dr. Abdulrahman O. Musaiger, Department of Food Sciences and Nutrition,
Faculty of Agricultural Sciences, United Arab Emirates University,
P.O. Box 17555, Al-Ain (United Arab Emirates)

Subject Index

Family doctor, infant feeding practices
152, 153
Fat
  absorption by infants 65, 66
  infant nutrition 59
  milk 64, 65
Fatty acids 61–63
  ω3 16–18
    preterm infants 16–23
  ω6 17
Feeding type, effect on development 11–16
Forced choice perferential looking tests 21
Fore milk 45
Formulas
  essential fatty acid composition 20
  nutritional components 2
  premature infants 58
Fruit juices 118, 120

Galactagogues 47
Galactopoiesis 39, 40, 48
  impairment 40, 41
Gastroenteritis 172, 173
Ghee, infant food 174

Hand feeding 49
Health workers, breastfeeding 179
Hind milk 45, 49
Hobart Child Welfare Association Report
  78
Home Inventory 11
Home stimulation environment 15
Hydrolysates, human milk fortifiers 66
Hyperlactation syndrome
  infant 49
  maternal 49–52

IgA, human milk 62
IgG, human milk 62
Illinois Psycholinguistic Abilities Test 9
Industry Code of Practice, Marketing of
  Infant Fomulas 101, 102
Infant(s)
  breastfed
    docosahexaenoic acid levels 18
    mental development 1, 2
  development, measurement 8, 9

feeding patterns 152, 153
formula 77
  distribution 185
  nutrition labeling 181
formula-fed
  docosahexaenoic acid levels 18
  mental development 1, 2, 121–126
  motor development 121, 125, 126
  physical growth 121–123
growth 121–123, 130, 131
human milk feeding 11–15
infant morbidity 132
influence of maids 177–179
mental development 121–126
morbidity 132
motor development 121, 125, 126
preterm
  effects of fatty acid deficiency 19–23
  human milk feeding 15, 16, 55–69
psychomotor development 7, 8
very low birth weight 19–23
visual maturation 19, 20
Insufficient milk syndrome 41, 45–49
Intelligence
  maternal 11
  measurement 9
Intrauterine growth 56

Korea, breastfeeding practices 114–126
Kuwait
  breastfeeding practices 168, 176, 177,
    179
  cessation 172

Lactalbumin hydrolysates 66
Lactation
  prenatal 35, 36
  problems 30, 31
Lactoengineering 66
Lactoferrin, human milk 62
Lactogenesis 36, 37
  impairment 38, 39
Lactose concentration, milk 67, 68
Language, measurement 9
Let down, see Milk ejection reflex
Linoleic acid 17, 18, 134
  visual maturation 19–23

Recall bias, surveys 90, 91
Reduction mammoplasty 36
Retinal development 17
Rickets 132
Rod photoreceptors 21, 22
Rotary evaporation, milk 67

Saudi Arabia
    breastfeeding practices 176, 177, 179
        cessation 172
Sheehan syndrome 39
Short Preschool Scale, Woodcock Test 9
Socioeconomic status
    breastfeeding practices 92–94
    weaning 119
Somatic growth 56, 65
Sonication 60, 61
South Australia, breastfeeding practices
    85, 88
Soy oil, effect on visual acuity 21–23
Staphylococcus
    aureus skin infection 43
    epidermidis 62
Stanford-Binet Intelligence Scale 9
Suckle 42, 43
Suckling 37, 41, 42
Suppressor peptides 40, 49

Tasmania, breastfeeding practices 85, 88
Teeth, infantile, appearance 121, 124
Tube feeding, milk 58, 59

Ultrasonic vibration 60
UNICEF 100
United Arab Emirates
    bottle feeding practices 169
    breastfeeding practices 169
Urbanization, breastfeeding 176, 177

Victoria, breastfeeding practices 85, 88, 89
Vietnamese immigrants, breastfeeding
    practices 95, 99
Visual acuity 19
    measurement 21
Visual evoked potentials 21
Visual-motor integration, measurement 9

Weaning
    foods 118–121
        nutrition labeling 181
    practices
        age levels 118–120
        Arabian Gulf countries 173–175,
            185, 186
        China 136, 137
Wechsler Adult Intelligence Scale 11
Western Australia, breastfeeding practices
    86–89
World Summit for Children 105